T0182036

Breeding Grasses and Protein Crops in the Era of Genomics

Gintaras Brazauskas · Gražina Statkevičiūtė
Kristina Jonavičienė
Editors

Breeding Grasses and Protein Crops in the Era of Genomics

Editors
Gintaras Brazauskas
Lithuanian Research Centre
  for Agriculture and Forestry (LAMMC)
Dotnuva
Lithuania

Kristina Jonavičienė
Lithuanian Research Centre
  for Agriculture and Forestry (LAMMC)
Dotnuva
Lithuania

Gražina Statkevičiūtė
Lithuanian Research Centre
  for Agriculture and Forestry (LAMMC)
Dotnuva
Lithuania

ISBN 978-3-030-07807-2   ISBN 978-3-319-89578-9 (eBook)
https://doi.org/10.1007/978-3-319-89578-9

Printed on acid-free paper

This Springer imprint is published by the registered company Springer International Publishing AG part of Springer Nature
The registered company address is: Gewerbestrasse 11, 6330 Cham, Switzerland

Preface

The joint meeting Breeding Grasses and Protein Crops in the Era of Genomics of Eucarpia Fodder Crops and Amenity Grasses Section and Protein Crops Working Group of Oil and Protein Crops Section took place in Vilnius, Lithuania, on September 11–14, 2017. A total of 142 participants from 24 countries attended the meeting hosted by the Lithuanian Research Centre for Agriculture and Forestry (LAMMC).

The meeting was subdivided into four sessions according to the thematic relevance to promote the discussion between both sections of Eucarpia. The first day had started with an opening session and the first scientific session Utilization of Genetic Resources and Pre-breeding, co-chaired by Roland Kölliker and Ulf Feuerstein. The key-note presentation was given by Odd Arne Rognli where he discussed utilization of genebank accessions to improve adaptation of perennial ryegrass to northern environmental conditions. The key-note presentation was then followed by seven further oral presentations and ten flash (max 5 min) presentations with subsequent "meet-and-greet" discussion during the coffee break. Session 2: Genetic Improvement of Quality and Agronomic Traits (co-chaired by Piet Arts and James P. Muir) opened the second day of the meeting. The key-note presentation was given by Bernadette Julier where she presented how the breeding of forage and grain legumes could contribute to increased protein self-sufficiency in the EU and China. Five further oral presentations were given in this session. Session 3: Breeding for Enhanced Stress Tolerance was co-chaired by Odd Arne Rognli and Diego Rubiales, while Torben Asp presented population genetics and GxE interaction analysis of the Eucarpia multi-site rust evaluation trial as a key-note presenter. Four oral presentations and eleven flash presentations followed by discussion concluded the second day of the meeting.

The third day was devoted to research visits to the Lithuanian Research Centre for Agriculture and Forestry (LAMMC), agri-business company Dotnuva Baltic, fertilizer plant Lifosa, and Feliksas Vaitelis organic beef cattle farm. Attendees of the meeting had possibility to see the breeding nurseries of LAMMC, seed production at Dotnuva Baltic, fertilizer production facilities at Lifosa and discuss farmers' perspective on forage and protein crops cultivated in Lithuania.

The fourth day of the meeting comprised Session 4: Implementation of Phenomic and Genomic Tools in Breeding. Two key-note presentations were given in this session. Kioumars Ghamkhar presented the use of LIDAR for forage yield measurement of perennial ryegrass field plots, while Steven Yates talked about the use of genetics and genomics to advance forage grass breeding. Two oral presentations and ten flash presentations concluded the meeting.

We gratefully acknowledge the efforts of the members of the scientific committee: Mario O. Aguilar, Paolo Annicchiarico, Torben Asp, Joost Baert, Philippe Barre, Susanne Barth, Didier Bazile, Beat Boller, Antonio M De Ron, Jean J. Drevon, Ulf Feuerstein, Deidré Fourie, Áslaug Helgadóttir, Christian Huyghe, Mike Humphreys, Hans J. Jacobsen, Bernadette Julier, Žydrė Kadžiulienė, Danutė Karčauskienė, Roland Kölliker, David Kopecký, Jan Nedělník, Liv Østrem, Roberto Papa, Izolda Pašakinskienė, Dirk Reheul, Isabel Roldán-Ruiz, Diego Rubiales, Tom Ruttink, Kadambot H. M. Siddique, Leif Skøt, Dejan Sokolović, Frederick Stoddard, Bruno Studer, Wojciech Święcicki, Lina Šarūnaitė, Vita Tilvikienė, Hari D. Upadhyaya, Bülent Uzun, Rajeev K. Varshney, Maria Carlota Vaz Patto, and Margarita Vishnyakova for the critical review of the offered papers. Their selfless efforts made it possible to have this meeting with the exciting scientific program and high-quality contributions to the proceedings of the meeting.

We also thank members of the local organizing committee: Agnė Jankauskienė, Andrius Aleliūnas, Ardas Kavaliauskas, Dangira Šidlauskienė, Eglė Norkevičienė, Giedrius Petrauskas, Irena Pabrinkienė, Kęstutis Tamošiūnas, Lina Šarūnaitė, Nijolė Lemežienė, Olakunle Kelvin Akinroluyo, Radvilė Nagrockaitė-Lelešienė, Vilma Kemešytė, Vita Tilvikienė, and Žydrė Kadžiulienė for making this meeting happen in its best form.

Finally, we thank all the participants for their active participation, desire to share their expertise, and discuss cutting-edge achievements of breeding grasses and protein crops in the era of genomics.

Gintaras Brazauskas
Gražina Statkevičiūtė
Kristina Jonavičienė

Contents

Utilization of Genetic Resources and Pre-breeding

**Utilization of Genebank Accessions to Improve Northern
Adaptation of Perennial Ryegrass (*Lolium perenne* L.)**   3
O. A. Rognli, R. Aavola, A. Aleliūnas, T. Asp, G. Brazauskas,
K. H. Gylstrøm, A. Helgadottir, M. Isolahti, M. R. Kovi,
T. A. Kristjánsdóttir, A. S. Larsen, P. Marum, C. Paina,
C. Persson, and S. Rancāne

**Wild and Primitive Common Bean Populations
in the Germplasm Collection at the MBG-CSIC**   9
A. M. De Ron, J. D. Pose, M. C. Menéndez-Sevillano,
and M. J. Ferreyra

**Patterns of Spring Growth and Phenology in Natural Populations
of *Lolium perenne* Under Contrasting Field Conditions**   14
A. M. Roschanski, P. Barre, A. Escobar-Gutiérrez, J. P. Sampoux,
H. Muylle, I. Thomas, K. J. Dehmer, and E. Willner

**Population Structure of Red Clover Ecotypes Collected
from Europe and Asia**   20
C. Jones, J. De Vega, D. Lloyd, M. Hegarty, S. Ayling, W. Powell,
and L. Skøt

**Enhancing the Sustainable Use of *Lolium perenne* Genetic
Resources from Genebanks in Plant Breeding and Research**   27
S. Bachmann-Pfabe, E. Willner, M. Oppermann,
S. Weise, and K. J. Dehmer

**Tetraploid *Festuca apennina* is Prone to Produce Triploid Hybrid
Progeny When Crossed with Diploid *Festuca pratensis***   33
B. Boller, T. Felder, and D. Kopecký

**Comparing Mitotic and Meiotic Tetraploid Red Clover Plants
for Seed Yield**   39
T. Vleugels, I. Roldán-Ruiz, and G. Cnops

**Domesticating Shrubby Native Legumes for Pastures
and Natural Grasslands**   46
J. P. Muir, F. S. Forrest, J. Brady, J. C. Dubeux, Jr., and W. D. Pitman

**Effects of Temperature on Growth During *in vitro* Embryo
Culture of Field Bean (*Vicia faba* var. *minor* L.)**   51
P. Ogrodowicz, M. Surma, T. Adamski, Z. Kaczmarek,
W. K. Święcicki, P. Stopyra, A. Kuczyńska, K. Krystkowiak,
and K. Mikołajczak

**Ecological and Agronomic Evaluation of *Dactylis glomerata*
subsp. *lobata***   56
E. Norkevičienė, A. Balsevičius, V. Kemešytė, B. Butkutė,
and G. Petrauskas

**Yield and Genetic Composition of Latvian ×*Festulolium* Cultivars
and Breeding Material**   62
P. Bērziņš, D. Ruņģis, S. Rancāne, A. Gailīte, V. Beļeviča, V. Stesele,
I. Vēzis, and A. Jansons

**Genetic Variability of the Most Important Traits in Alfalfa
Cultivars Under Lithuanian Conditions**   67
A. Liatukienė

Tetraploid Induction in *Lolium multiflorum*   73
O. Akinroluyo, G. Statkevičiūtė, and V. Kemešytė

**The Sense of Testing Sheep Preference in Tall Fescue
Variety Evaluation**   78
M. Cougnon, J. Baert, and D. Reheul

Chemical Composition of Zigzag Clover (*Trifolium medium* L.)   83
B. Butkutė, N. Lemežienė, A. Padarauskas, E. Norkevičienė,
and L. Taujenis

**Wild and Semi Natural Ecotypes of Perennial Grasses
and Legumes – for Breeding Purposes**   88
N. Lemežienė, V. Stukonis, V. Kemešytė, and E. Norkevičienė

Breeding for Improved Seed Yield of Red Clover   96
G. Petrauskas, J. Mikaliūnienė, E. Norkevičienė, G. Statkevičiūtė,
and V. Kemešytė

Genetic Improvement of Quality and Agronomic Traits

**Breeding Forage and Grain Legumes to Increase EU's
and China's Protein Self-sufficiency** 103
B. Julier, L. Skøt, S. Weise, Đ. Karagić, I. Roldán-Ruiz, P. Barre,
and D. Lloyd

**Identification of Loci Controlling Timing of Stem Elongation
in Red Clover Using GBS of Pooled Phenotypic Extremes** 109
Å. Ergon, Ø. W. Milvang, O. Milvang, and M. R. Kovi

**Grain Yield and Morphology of Dwarf *vs* Tall White Lupin
in Mediterranean Environments** 113
P. Annicchiarico, A. M. Carroni, P. Manunza, C. Huyghe,
and L. Pecetti

**Deciphering Grain Legumes Quality Riddle: The Genomics
of Bioactive Compounds** 118
M. C. Vaz Patto, E. Mecha, A. B. Pereira, S. T. Leitão, M. L. Alves,
and M. R. Bronze

**Manipulation of Rumen Nitrogen Flows by Use of Festulolium
Hybrids in Rumen Simulation Fermenters** 121
S. Kamau, A. Belanche, M. Humphreys, and A. Kingston-Smith

**Paternity Test for Forage Yield Improvement in Timothy
Polycross Breeding** 125
T. Tanaka, K. Tamura, K. Ashikaga, H. Fujii, and T. Yamada

**Preliminary Estimation of Variation of Alkaloids Content
in White Lupin (*Lupinus albus* L.) Collection** 131
W. Rybiński, M. Kroc, W. Święcicki, P. Wilczura, K. Kamel,
P. Barzyk, and W. Mikulski

**Yields and Quality of Some Perennial Legumes
in the Czech Republic** 137
D. Knotová, J. Pelikán, J. Skládanka, and P. Knot

Breeding for Enhanced Stress Tolerance

**Low Crown Rust Resistance in Norwegian Material of *Lolium
perenne* and ×*Festulolium*** 145
L. Østrem, T. Asp, M. Ghesquière, Y. Sanada, and O. A. Rognli

**Unravelling the Genetic Control of Bacterial Wilt Resistance
in Ryegrass: Achievements, Prospects and Challenges** 150
R. Kölliker, V. Knorst, L. Hersemann, F. Widmer, and B. Studer

**Genotyping of Festulolium Cultivars Involved in EUCARPIA
Multi-site Trial Using DArT Markers and GISH** 155
D. Kopecký, J. Baert, S. Barth, J. Bartoš, V. Černoch, J. Doležel,
D. Grogan, J. Harper, M. Humphreys, T. Książczyk, L. Østrem,
E. Paszkowski, D. Sokolovič, Z. Zwierzykowski, and M. Ghesquière

**Transcriptomic Analysis of Drought-Sensitive and Tolerant
Genotypes of Perennial Ryegrass (_Lolium perenne_ L.)** 160
M. O. Leyva, I. Nagy, S. Hentrup, S. Byrne, and T. Asp

Selection of Perennial Ryegrass Spaced Plants in Sward Conditions . . . 166
J. Aper, A. Ghesquiere, and J. Baert

**Variation in Intrinsic Water Use Efficiency Between Perennial
Ryegrass Genotypes Differing for Drought Tolerance** 171
P. Westermeier, R. Schäufele, and S. Hartmann

**Physiological Indicators of Tolerance to Soil Water Deficit
in _Lolium multiflorum/Festuca arundinacea_ Introgression Forms** 176
K. Masajada, A. Augustyniak, D. Perlikowski, D. Ratajczak,
W. Zwierzykowski, I. Pawłowicz, and A. Kosmala

**Incidence of Six Grass Species by _Fusarium_ sp. as a Cause
of Silvertop** 181
O. Trněný and J. Nedělník

**Crown Rust and Brown Blight Infection of Perennial Ryegrass
Cultivars Representing Three Maturity Groups** 186
R. Aavola, P. Soovāli, S. Tamm, and A. Bender

**Correlation of Seed Yield and Its Components and Chlorophyll
Fluorescence Parameters in the Narrow Leafed Lupin
(_Lupinus angustifolius_ L.)** 191
B. Górynowicz, W. Święcicki, W. Pilarczyk, and W. Mikulski

Seventh Workshop of the EUCARPIA Festulolium Working Group . . . 196
M. Ghesquière

Implementation of Phenomic and Genomic Tools in Breeding

**Using LIDAR for Forage Yield Measurement of Perennial Ryegrass
(_Lolium perenne_ L.) Field Plots** 203
K. Ghamkhar, K. Irie, M. Hagedorn, J. Hsiao, J. Fourie, S. Gebbie,
C. Flay, B. Barrett, A. Stewart, and A. Werner

**An Indoor Screening Method for Reduced Fall Dormancy
in Alfalfa** 209
A. Bertrand, A. Claessens, and S. Rocher

Application of Multivariate Analysis for Genotype Evaluation in Soybean 215
V. Perić, M. Srebrić, A. Nikolić, and S. Mladenović-Drinić

Polymorphism of Soybean Cultivars and Breeding Lines Revealed by Marker *Satt100* Associated with the E_7 Locus 220
D. Zharikova, S. Ivanyuk, G. Chebotar, O. Korniychuk, and S. Chebotar

Genetic Diversity Assessment of Two *Medicago sativa* Genes: CAD and WXP1 226
C. Gréard, P. Barre, S. Flajoulot, S. Santoni, and B. Julier

Testing for Freezing Tolerance in Perennial Ryegrass 231
A. Aleliūnas and G. Brazauskas

The Multifunctional Role of Legumes in Vineyards and Orchards 237
T. Vymyslický, J. Lang, and D. Knotová

Hydroponic Evaluation of Growth and Nutrient Uptake in a *Lolium/Festuca* Introgression Series 243
D. Gasior, J. Harper, A. Thomas, C. Evans, R. Mathews, D. Allen, M. Humphreys, and I. Armstead

Use of GBS for Lucerne Variety Distinction 249
B. Julier, P. Barre, P. Lambroni, S. Delaunay, F. Lafaillette, M. Thomasset, and V. Gensollen

Genome-Wide and Functional Gene Analysis of Perennial Ryegrass for Improved Growth Under Water Limiting Conditions (GrowGene) 254
K. Jonavičienė, G. Statkevičiūtė, and A. Aleliūnas

Genomic Selection for Biomass Yield of Perennial and Annual Legumes 259
P. Annicchiarico, N. Nazzicari, L. Pecetti, and M. Romani

Author Index 265

Utilization of Genetic Resources
and Pre-breeding

Utilization of Genebank Accessions to Improve Northern Adaptation of Perennial Ryegrass (*Lolium perenne* L.)

O. A. Rognli[1][✉], R. Aavola[2], A. Aleliūnas[3], T. Asp[4],
G. Brazauskas[3], K. H. Gylstrøm[5], A. Helgadottir[6], M. Isolahti[7],
M. R. Kovi[1], T. A. Kristjánsdóttir[6], A. S. Larsen[8], P. Marum[5],
C. Paina[4], C. Persson[9], and S. Rancāne[10]

[1] Department of Plant Sciences, Faculty of Biosciences,
Norwegian University of Life Sciences, Ås, Norway
odd-arne.rognli@nmbu.no
[2] Estonian Crop Research Institute, Jõgeva, Estonia
[3] Institute of Agriculture, Lithuanian Research Centre for Agriculture
and Forestry, Kėdainiai reg., Lithuania
[4] Department of Molecular Biology and Genetics,
Aarhus University, Slagelse, Denmark
[5] Graminor AS, Ridabu, Norway
[6] Department of Land and Animal Resources,
Agricultural University of Iceland, Reykjavík, Iceland
[7] Boreal Plant Breeding Ltd., Jokioinen, Finland
[8] DLF Research Center, Store Heddinge, Denmark
[9] Lantmännen ek. för., Svalöv, Sweden
[10] Latvia University of Agriculture, Jelgava, Latvia

Abstract. Perennial ryegrass (*Lolium perenne* L.) is increasingly important as a forage grass in the Nordic and Baltic region, especially in the light of climate change. However, winter hardiness and persistence need to be improved. This article describes the long-term Nordic/Baltic public-private partnership for pre-breeding in perennial ryegrass, started in 2012. The basis for the project is detailed genotyping and phenotyping of nearly 400 genebank accessions, followed by generation of new broad-based diploid and tetraploid breeding populations, locally adapted and trait-based populations. Preliminary results confirm the need for more robust and stable cultivars with wider adaptations, and genebank accessions need to be introgressed and recombined with adapted materials followed by natural/artificial selection in order to create novel germplasm for these regions.

Keywords: *Lolium perenne* · Pre-breeding · Genebank · Accessions
Phenotyping · Genotyping

© Springer International Publishing AG, part of Springer Nature 2018
G. Brazauskas et al. (Eds.): *Breeding Grasses and Protein Crops in the Era of Genomics*, pp. 3–8, 2018.
https://doi.org/10.1007/978-3-319-89578-9_1

1 Introduction

Perennial ryegrass (*Lolium perenne* L.) has superior feed quality and productivity, and it is especially well adapted to the coastal climate of Western Europe (Humphreys et al. 2010). In Northern Europe, the expected climate changes will most likely improve the conditions for forage production because the growing season will be longer (1–3 months) with milder and rainier autumns and winters (Olesen et al. 2011). Thus, we expect perennial ryegrass to be grown commercially further north and east. However, the main challenges for perennial ryegrass is its susceptibility to low-temperature pathogens, and inadequate growth cessation in the autumn to allow for sufficient cold hardening and winter survival, with low persistency as the result (Solberg et al. 1994; Østrem et al. 2015). This will be exacerbated by the higher temperatures in the new climate changing the temperature-photoperiod regimes affecting plant phenology, winter survival and seasonal yield distribution. The genetic diversity available in the current Nordic/Baltic perennial ryegrass germplasm is probably restricted since the species is not native to the northern and continental regions (Rognli et al. 2013). Therefore, exotic materials should be identified, introgressed and recombined with existing breeding populations to serve as new genetic resources for development of cultivars for the future climate in these regions. This article describes the structure and activities of the Nordic Public-Private Partnership (PPP) on pre-breeding (Nilsson et al. 2016) of perennial ryegrass. The project aims at improving winter hardiness, persistence and other important traits for northern Europe by: (i) investigating the current adaptation potential of commercial cultivars; (ii) collecting and documenting plant material with large genetic variation from several parts of the world; (iii) creating breeding populations with large variation and populations selected for extremes of different traits; and (iv) developing tools for using genomic based selection in further breeding programmes.

2 Materials and Methods

2.1 Plant Material, Phenotyping and Pre-breeding

Breeding perennial forage grasses is a long-term activity taking 10–15 years until cultivars are released. This pre-breeding project was designed to run from 2012 to 2020, and the project started by obtaining seed samples of 393 accessions from gene banks worldwide (Table 1). These accessions were seed multiplied to obtain enough seed for regular field trials, and at the same time established as replicated single plant experiments for detailed phenotyping in Denmark (DK), Sweden (SE), Norway (NO) and Finland (FI). Based on this phenotyping, 22 contrasting trait-based local populations have been composed, Syn-1 and Syn-2 seeds are produced, and field testing of these will be established in 2019. In addition, the accessions were sown as single rows in DK for proper determination of heading date in dense stand.

A broad-based diploid breeding population was constructed based on 350 populations using 10 plants from each population completely randomized within each of two crossing blocks, one in DK and one in NO. Crossing blocks in different

Table 1. Geographic origin and types of gene bank accessions used in this project.

Country	Wild	Cultivars/breeding lines	Unknown	Total
Austria		1	1	2
Belgium		1		1
Canada			1	1
Czech Rep	9	2	5	16
Denmark		27	1	28
Estonia	2	7		9
Finland		4	1	5
France	33	5		38
Germany	32	7		39
Hungary	12	2	2	16
Ireland		2		2
Italy	6			6
Japan		3		3
Kyrgyzstan[a]		1		1
Latvia	2	15	2	19
Lithuania	16	15		31
Moldova	1			1
Norway	10	13		23
Poland	29	11	2	42
Romania	15	1		16
Russia	2	7	3	12
Switzerland	8	3		11
Slovakia	2	1		3
Slovenia		1		1
Sweden	11	17	2	30
The Netherlands		6		6
Turkey	9			9
UK	5	6		11
Ukraine	4		2	6
USA		2	1	3
Unknown			2	2
Total	208	160	25	393

[a]Later confirmed to be *L. multiflorum* based on genotyping.

environments were used to reduce the effect of selection due to differences in response to photoperiod and other climatic factors. One ear per plant was harvested and used to establish one plant per ear for production of the 2nd generation seed. Half of each crossing block was harvested early, the other late, and balanced bulks constructed.

In 2016, the broad-based breeding population was sown on large plots at 1–2 locations per country (a total number of 13 plots) for natural selection to create locally adapted germplasms. In addition, new tetraploid genotypes were made by chromosome doubling of 1000 plants of the broad-based population, these are being intercrossed in NO and DK to create a new broad-based tetraploid breeding population using similar procedures as with the diploid population. Detailed phenotyping of 250 tetraploids is performed by the Latvian (LV) partner.

Field trials in five countries (Iceland (IS), SE, NO, Estonia (EE) and FI) for testing winter hardiness and phenotypic stability of 22 perennial ryegrass cultivars was completed (Helgadottir et al. 2018), and regional-specific populations (5 diploid, 5 tetraploid) created based on crossing surviving plants of diploid and tetraploid cultivars at each location (Syn-1 and Syn-2 seed produced by 2019). Forage quality of all cuts of the 2nd harvest year from the field trials in NO, IS, FI and EE has been analysed. In addition the freezing tolerance (LT_{50}) of the cultivars was determined in artificial freezing tests by the Lithuanian partner. In 2014, field experiments with seeds multiplied from the original accessions were established in seven countries (IS, SE, NO, FI, DK, EE and Lithuania (LT)) for testing using regular plot size. A core set of 138 accessions is present at all locations, and yield and other traits have been recorded for 2–3 years (2015–2017). In addition, forage quality have been measured in FI and DK using NIRS. An artificial freezing test of 150 accessions, including the core set, have been performed by the Lithuanian partner (Aleliunas et al. 2017).

Partners in the project and sites for field trials are depicted in Fig. 1.

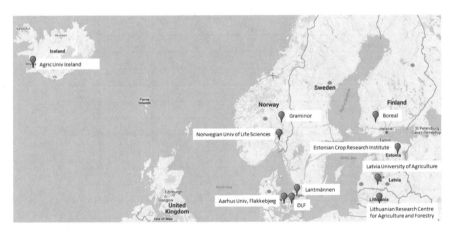

Fig. 1. Nordic and Baltic partners in the PPP perennial ryegrass project indicate field testing sites in addition to the main sites of the partners.

2.2 Genotyping, Ploidy Testing

Pooled leaf samples from plants of the original seed lots of the 380 gene bank accessions (Table 1) were genotyped by sequencing (GBS) to estimate genome wide SNP allele frequencies (see Byrne et al. 2013). Similar genotyping has been or is being performed on the 22 cultivars tested in field trials, the 10 location-specific 2x and 4x populations created from surviving plants of these cultivars (syn-1 and syn-2), and the 22 diverse trait-based local populations created from the single plant phenotyping trials (syn-1, syn-2 and the parents). The latter genotyping data is being used to study the differential parental contributions when synthetic populations are generated by open pollination. In addition, leaf samples of about 6,700 single plants have been collected from the single plant phenotyping trials (NO, SE and FI) and stored for genotyping in a later phase of the project. These genomic resources are powerful resources for association studies and eventually, development of genomic selection schemes. It is important for characterization and further use of gene bank accessions to have exact knowledge about the ploidy level of each accession. Thus, a full flow-cytometric analysis of all accessions was carried out in 2015.

3 Results and Discussion

The results from the multi-site field trials confirm the need for more robust and stable cultivars with wider adaptations (Helgadottir et al. 2018). Cultivars, locations and years showed large interactions, and local cultivars were generally better adapted than cultivars from Central Europe. Diploid cultivars were consistently more frost tolerant than tetraploid cultivars. The Norwegian cultivar 'Falk', which was found to be a diploid/tetraploid mixture (about 50/50) by flow-cytometry, turned out to be the most stable cultivar across environments. This is interesting and indicate that using ploidy-mixtures could be a strategy to improve persistency of perennial ryegrass grown in these northern regions.

The preliminary results from the multi-site field testing of accessions indicate that none of the accessions is better than the best locally adapted cultivar. Thus, it seems like our initial assumption that introgression and recombination of exotic materials followed by natural/artificial selection is needed to create novel germplasm. Similar results were reported by Ghesquiere and Baert (2013) when they evaluated families from pair-crosses between elite breeding materials and 30 core-collection accessions. Also, most of the accessions seem to be susceptible to rust attack. Freezing tests of accessions confirmed the results from freezing tests of the cultivars, i.e. diploids are consistently more freezing tolerant than tetraploids (Aleliunas et al. 2017).

Flow-cytometric analyses of the ploidy level of the accessions showed that as much as about 16% of the accessions were tetraploids while they were classified as diploids in the genebank.

Acknowledgements. The PPP prebreeding in perennial ryegrass project is funded by the Nordic Ministers through Nordgen, and by the Nordic breeding companies involved. The funding of the research activities of the Baltic partners is provided nationally, e.g. the Research Council of Lithuania (Grant No. MIP-64/2015).

References

Aleliūnas A, Paina C, Statkevičiūtė G, Asp T, Brazauskas G (2017) Population-based genome-wide association analysis of freezing tolerance in perennial ryegrass. Submitted to the PloS ONE

Byrne S, Czaban A, Studer B, Panitz F, Bendixen C et al (2013) Genome wide allele frequency fingerprints (GWAFFs) of populations via genotyping by sequencing. PLoS One 8(3):e57438. https://doi.org/10.1371/journal.pone.0057438

Ghesquiere A, Baert J (2013) The use of genebank accessions in the breeding programme of *Lolium perenne*. In: Barth S, Milbourne D (eds) Breeding strategies for sustainable forage and turf grass improvement. Springer, Netherlands, pp 295–299

Helgadóttir Á, Aavola R, Isolahti M, Marum P, Persson C, Aleliunas A, Brazauskas G, Krisjánsdóttir TA, Asp T, Rognli OA (2018) Adaptability and phenotypic stability of Lolium perenne L. cultivars of diverse origin grown at the margin of the species distribution. J Agron Crop Sci https://doi.org/10.1111/jac.12273

Humphreys M, Feuerstein U, Vandewalle M, Baert J (2010) Ryegrasses. In: Boller B, Posselt U, Veronesi F (eds) Fodder crops and amenity grasses. handbook of plant breeding, vol 5. Springer, New York, pp 211–260. https://doi.org/10.1007/978-1-4419-0760-8_10

Olesen JE, Trnka M, Kersebaum KC, Skjelvåg AO, Seguin B, Peltonen-Sainio P, Rossi F, Kozyra J, Micale F (2011) Impacts and adaptation of European crop production systems to climate change. Eur J Agron 34:96–112

Nilsson A, von Bothmer R, Johanesson T, Nybom H, Bendevis MA, Bengtsson T, Rognli OA (2016) Promoting Nordic plant breeding for the future. In: Hägnefelt A. (ed) PPP public private partnership for pre-breeding. Nordic Council of Ministers, Copenhagen, ANP 2016:730, p 27. http://dx.doi.org/10.6027/ANP2016-730

Rognli OA, Fjellheim S, Pecetti L, Boller B (2013) Semi-natural grasslands as a source of genetic diversity. Grassland Sci Eur 18:303–313

Solberg ET, Rognli OA, Østrem L (1994) Potential for improving adaptation of Lolium perenne L. to continental climates in Norway. Euphytica 77:183–191

Østrem L, Rapacz M, Larsen A, Dalmannsdottir S, Jørgensen M (2015) Influences of growth cessation and photoacclimation on winter survival of non-native Lolium-Festuca grasses in high-latitude regions. Environ Exp Bot 111:21–31

Wild and Primitive Common Bean Populations in the Germplasm Collection at the MBG-CSIC

A. M. De Ron[1(✉)], J. D. Pose[1], M. C. Menéndez-Sevillano[2],
and M. J. Ferreyra[2]

[1] Biology of Agrosystems, Misión Biológica de Galicia (MBG),
Spanish National Research Council (CSIC), Pontevedra, Spain
amderon@mbg.csic.es
[2] Active Bank of Northwestern Argentina (BANOA), EEA Salta,
National Agricultural Technology Institute (INTA), Salta, Argentina

Abstract. In the Misión Biológica de Galicia, Spanish National Research Council (MBG-CSIC) there is a germplasm collection of common bean (*Phaseolus vulgaris* L.) that includes wild and cultivated accessions. In the case of the cultivated populations from the Andean Region, some of them are considered "primitive" since they are unimproved landraces. The aim of this study was to describe some characteristics of the wild and primitive populations and their area of origin.

Keywords: Domestication · *Phaseolus vulgaris* · Phenotypic traits
Variability

1 Introduction

The common bean is the most important grain legume for direct human consumption on a global scale. Among the main food crops, the common bean shows the greatest variation in growth habit, seed characteristics (size, shape and colour) and maturation time. This variability enables its production in a wide range of cropping systems and agrosystems in different continents. Andean domesticated common bean germplasm is remarkably diverse in plant and seed morphology and agroecological adaptation, but it has a narrow genetic base according to the molecular analysis (Beebe et al. 2001) as compared to Mesoamerican germplasm.

Northwestern Argentina (NOA) has been considered as an unique segment of the Andean gene pool, isolated from other bean germplasm since and intermediate or weedy types have been described in this area (Beebe et al. 2001). In this area there is still a considerable amount of genetic diversity in many wild populations and also in the primitive cultivated landraces since traditional agriculture in small farms is still practised in isolated valleys that include the use of primitive or unimproved landraces of "porotos" or beans often intercropped with maize (Parodi 1953). The wild forms grow in Argentina through the provinces of Salta, Jujuy and Tucumán, along the valleys of the Eastern Andean Mountain Range.

The aim of this study was to describe some characteristics of the wild and primitive common bean populations and their origin in the Andean region.

© Springer International Publishing AG, part of Springer Nature 2018
G. Brazauskas et al. (Eds.): *Breeding Grasses and Protein Crops in the Era of Genomics*, pp. 9–13, 2018.
https://doi.org/10.1007/978-3-319-89578-9_2

2 Material and Methods

Plant material

The common bean collection at the MBG-CSIC includes 2014 accessions from Europe (17 countries), The Americas (15), Asia (4), Africa (1) and Oceania (1). This collection incorporated wild and primitive populations from the NOA in South America that is one of the domestication centers (together with Mesoamerica) of the species (Singh et al. 1991).

3 Results and Discussion

The Table 1 shows the origin of the wild and primitive populations maintained in the germplasm collection at the MBG-CSIC.

Table 1. Origin of the wild and primitive populations maintained in the germplasm collection at the MBG-CSIC.

Code	Code2	Type	Location	Province	Altitude (masl)	Year
PHA-0725	VAV6388	Wild	Chicoana	Salta	1450	1991
PHA-0726	VAV6389	Wild	Chicoana	Salta	1400	1991
PHA-0727	VAV6368	Wild	Trancas	Tucumán	2600	1986
PHA-0728	VAV6395	Wild	Chicoana	Salta	1400	1993
PHA-0729	VAV6382	Wild	Salta	Salta	1800	1991
PHA-0730	VAV6396	Wild	La Caldera	Salta	1300	1993
PHA-0731	VAV6385	Wild	Chicoana	Salta	1450	1991
PHA-0732	VAV6387	Wild	Chicoana	Salta	1400	1991
PHA-0733	VAV6397	Wild	Santa Victoria	Salta	1630	1993
PHA-0733	VAV6406	Wild	Salta	Salta	1690	1993
PHA-0734	MCM168	Wild	San Carlos	Salta	1600	1994
PHA-0736	VAV6398	Wild	Valle Grande	Jujuy	1470	1993
PHA-0737	VAV6362	Wild	Tumbaya (route)	Jujuy	1850	1986
PHA-0738	VAV6363	Wild	Tiraxi (route)	Jujuy	1670	1986
PHA-0739	VAV6364	Wild	Santa Victoria	Salta	2340	1986
PHA-0740	VAV6365	Wild	Santa Victoria	Salta	2600	1986
PHA-0741	VAV6369	Wild	Trancas	Tucumán	1310	1986
PHA-0742	VAV6370	Wild	Trancas	Tucumán	1520	1986
PHA-0743	MCM69	Primitive	Iruya	Salta	2900	1986
PHA-0744	MCM132	Primitive	Iruya	Salta	2900	1986
PHA-0746	VAV5674	Primitive	Tilcara	Jujuy	2400	1977
PHA-0747	VAV5873	Primitive	Tilcara	Jujuy	2400	1978
PHA-0748	VAV5868	Primitive	Santa Victoria	Salta	2400	1978
PHA-0749	VAV5882	Primitive	Santa Victoria	Salta	2400	1978

(continued)

Table 1. (*continued*)

Code	Code2	Type	Location	Province	Altitude (masl)	Year
PHA-0750	VAV6198	Primitive	Santa Victoria	Salta	2400	1981
PHA-0752	VAV5867	Primitive	Santa Victoria	Salta	2400	1978
PHA-0753	MCM121	Primitive	Santa Victoria	Salta	2400	1986
PHA-0754	MCM123	Primitive	Santa Victoria	Salta	2400	1986
PHA-0755	MCM122	Primitive	Santa Victoria	Salta	2400	1986
PHA-0756	MCM128	Primitive	Santa Victoria	Salta	2400	1986
PHA-0757	MCM109	Primitive	Iruya	Salta	3200	1986
PHA-0758	MCM116	Primitive	Iruya	Salta	2800	1986
PHA-0759	MCM110	Primitive	Iruya	Salta	3020	1986
PHA-0760	MCM130	Primitive	Iruya	Salta	2800	1986
PHA-0761	MCM79A	Primitive	Santa Victoria	Salta	2400	1986
PHA-0762	MCM32	Primitive	Santa Victoria	Salta	2600	1986
PHA-0763	MCM75	Primitive	Santa Victoria	Salta	2400	1986
PHA-0764	MCM14	Primitive	Santa Victoria	Salta	2400	1986
PHA-0765	MCM85	Primitive	Santa Victoria	Salta	2400	1986
PHA-0766	MCM44	Primitive	Santa Victoria	Salta	2700	1986
PHA-0767	MCM88	Primitive	Santa Victoria	Salta	2400	1986
PHA-0768	MCM15	Primitive	Santa Victoria	Salta	2400	1986
PHA-0769	MCM2	Primitive	Iruya	Salta	2900	1986
PHA-0770	MCM78	Primitive	Santa Victoria	Salta	2400	1986
PHA-0772	VAV5883	Primitive	Santa Victoria	Salta	2500	1978
PHA-0773	VAV5878	Primitive	Tilcara	Jujuy	2400	1978
PHA-0774	VAV5874	Primitive	Tilcara	Jujuy	2400	1978
PHA-0775	VAV5881	Primitive	Santa Victoria	Salta	2400	1978
PHA-0777	MCM90	Primitive	Santa Victoria	Salta	2400	1986
PHA-0780	VAV3716	Primitive	Iruya	Salta	2900	1971
PHA-0782	VAV5672	Primitive	Tilcara	Jujuy	2400	1973
PHA-0852	MCM18	Primitive	Santa Victoria	Salta	2400	1986
PHA-0855	MCM194	Primitive	Santa Victoria	Salta	2600	1996
PHA-0856	MCM201	Wild	Santa Victoria	Salta	2500	1996
PHA-0858	MCM207	Wild	Chicoana	Salta	1540	1996
PHA-0860	VAV3733	Primitive	Tilcara	Jujuy	2400	1972
PHA-0867	VAV5675	Primitive	Santa Victoria	Salta	2400	1977
PHA-0875	VAV6366	Wild	Jujuy	Jujuy	1720	1986
PHA-0876	VAV6367	Wild	Jujuy	Jujuy	1650	1986
PHA-0878	VAV6381	Wild	Jujuy	Salta	1800	1991
PHA-0880	MCM87	Primitive	Santa Victoria	Salta	2400	1986
PHA-0882	MCM1	Primitive	Iruya	Salta	2900	1986
PHA-0883	MCM7	Primitive	Iruya	Salta	2900	1986
PHA-0884	MCM9	Primitive	Iruya	Salta	2900	1986

(*continued*)

Table 1. (*continued*)

Code	Code2	Type	Location	Province	Altitude (masl)	Year
PHA-0885	MCM16	Primitive	Santa Victoria	Salta	2400	1986
PHA-0886	MCM26	Primitive	Santa Victoria	Salta	2400	1986
PHA-0887	MCM73	Primitive	Santa Victoria	Salta	2400	1986
PHA-0888	MCM82	Primitive	Iruya	Salta	2900	1986
PHA-0889	MCM84	Primitive	Iruya	Salta	2900	1986
PHA-0890	MCM89	Primitive	Santa Victoria	Salta	2400	1986
PHA-0891	MCM99	Primitive	Santa Victoria	Salta	2400	1986
PHA-0892	MCM100	Primitive	Santa Victoria	Salta	2400	1986
PHA-0893	MCM131	Primitive	Iruya	Salta	2900	1986
PHA-0894	MCM133	Primitive	Iruya	Salta	2900	1986
PHA-0895	MCM199	Primitive	Tumbaya	Jujuy	1900	1996
PHA-0896	MCM200	Wild	Tumbaya	Jujuy	1900	1996
PHA-0897	MCM202	Wild	Tumbaya	Jujuy	1900	1996
PHA-0898	MCM206	Wild	Chicoana	Salta	1400	1996
PHA-1400	VAV6404	Primitive	Los Toldos	Salta	1600	1993

Code - MBG-CSIC accession name, *Code2* - original accession name.

Wild beans of South America are very small seeded compared with their corresponding domesticated forms. Common bean has evolved during domestication from small- to large-seeded forms, from extreme indeterminate climbing types to determinate bush types, from seed dormancy and water impermeability of the seed coat to lack of dormancy and water permeable seed coat and from highly fibrous pod wall and shattering forms to lack of fibres and no shattering types. There was also a loss of seed coat pigmentation in some populations and a reduction of the content in toxic materials, which have probably improved the flavour and caused the loss of seed protection against predators (Smartt 1988).

The missions for collect the wild bean populations from the NOA included in the present study were carried out since 1971 in the Argentinean provinces of Jujuy, Salta and Tucumán. Mostly of the studied populations grow in the Yungas ecoregion, a narrow strip in the slopes of the Andean mountains in the provinces of Salta, Jujuy, Tucumán and Catamarca reaching heights of up to 3000 masl. The climate is warm and humid with summer rains and frost in winter, annual rainfall reaches 2500 mm. It is a region of mountains, canyons and plains with forest acid soils with its three characteristic horizons: humifer, topsoil and mineral earth. The wild and primitive bean germplasm maintained at the MBG-CSIC was studied for their phenotypic (Menéndez-Sevillano 2002; De Ron et al. 2004; Santalla et al. 2004) and genetic variation (Galván et al. 2006).

4   Conclusions

As a conclusion, an efficient use of the genetic resources requires an understanding of the origin and the variation in the germplasm collections available. Variation in some quantitative traits revealing agronomic value together with other associated with regional preferences, i.e. seed coat colour and pattern, is relevant for breeding purposes. The use of wild and primitive populations (with low inputs for their growth) in breeding programs (Singh et al. 1995) along with improved varieties will allow increasing the sustainability of the bean crop. These common bean germplasm is being conserved *in situ;* the wild populations in their ecosystems and the domesticated ones in farms through a participatory program with farmers of the NOA under traditional growing methods.

Acknowledgments. This study was funded through the projects 2016SU0004 (CSIC, Spain) and RFP2015-0008 (INIA, Spain).

References

Beebe S, Rengifo J, Gaitan E, Duque MC, Tohme J (2001) Diversity and origin of Andean landraces of common bean. Crop Sci 41:854–862

De Ron AM, Menéndez-Sevillano MC, Santalla M (2004) Variation in primitive landraces of common bean (*Phaseolus vulgaris* L.) from Argentina. Genet Resour Crop Evol 51:883–894

Galván MZ, Menéndez-Sevillano MC, De Ron AM, Santalla M, Balatti PA (2006) Genetic diversity among wild common beans from Northwestern Argentina based on morphoagronomic and RAPD data. Genet Resour Crop Evol 53:891–900

Menéndez-Sevillano MC (2002) Estudio y conservación del germoplasma silvestre y primitivo de *Phaseolus vulgaris* L. en el Noroeste de Argentina. Ph.D. Thesis. University of Santiago de Compostela, Spain

Parodi LR (1953) Relaciones de la agricultura prehispánica con la agricultura actual. Anales de la Academia Nacional de Agronomía y Veterinaria de Buenos Aires 1:115–167

Santalla M, Menéndez-Sevillano MC, Monteagudo AB, De Ron AM (2004) Genetic diversity of Argentinean common bean and its evolution during domestication. Euphytica 135:75–87

Singh SP, Gepts P, Debouck D (1991) Races of common bean (*Phaseolus vulgaris*, Fabaceae). Econ Bot 45:379–396

Singh SP, Molina A, Gepts P (1995) Potential of wild common bean for seed yield improvement of cultivars in the tropics. Can J Plant Sci 75:807–813

Smartt J (1988) Morphological, physiological and biochymical changes in *Phaseolus* beans under domestication. In: Gepts P (ed) Genetics resources of *Phaseolus* beans: their maintenance, domestication, evolution and utilization. Kluwer, Dordrecht, pp 543–560

Patterns of Spring Growth and Phenology in Natural Populations of *Lolium perenne* Under Contrasting Field Conditions

A. M. Roschanski[1(✉)], P. Barre[2], A. Escobar-Gutiérrez[2],
J. P. Sampoux[2], H. Muylle[3], I. Thomas[4], K. J. Dehmer[1],
and E. Willner[1]

[1] Leibniz Institute of Plant Genetics and Crop Plant Research (IPK),
Inselstr. 9, 23999 Malchow/Poel, Germany
roschanski@ipk-gatersleben.de
[2] INRA, Centre Nouvelle-Aquitaine-Poitiers,
UR4 (UR P3F), 86600 Lusignan, France
[3] ILVO, Plant Genetics and Breeding Section,
Caritasstraat 21 B, 9090 Melle, Belgium
[4] IBERS-Aberystwyth University,
Plas Goggerdan, Aberystwyth, Ceredigion, UK

Abstract. The ecotypic diversity of perennial ryegrass (*Lolium perenne* L.) is a major genetic resource for breeding programs. In three replicated micro-sward trials in France, Belgium and Germany, we measured spring growth and recorded heading date of round 400 genebank accessions from the natural diversity of *L. perenne* that were selected as to represent the wide range of variability in this species. We observed marked differences between trial locations as well as interaction between accessions and locations in the timing of spring growth rates along growing-degree-days (GDDs). These preliminary results are part of a wider project aiming to investigate the natural adaptation of perennial ryegrass to various regional climates across its spontaneous area of presence in Europe.

Keywords: Ecotype · Heading date · Growth rate · *Lolium perenne*
Natural diversity · Spring growth · Thermal time

1 Introduction

Grasslands represent the most widespread agricultural land use in Europe. Among grassland species, *Lolium perenne* is one of the most prevalent and economically important grass species (Humphreys et al. 2010, Sampoux et al. 2013). *Lolium perenne* has undergone spontaneous in-situ differentiation into ecotypes (Boller and Greene 2010), presumably adapted to local environmental conditions. A large number of natural *L. perenne* populations are maintained in European genebanks as plant genetic resources (PGRs). With the concern of ongoing climate change, the introduction of ecotypic variability into breeding programs may contribute to develop cultivars with high ecophysiological resilience (Boller and Greene 2010), while meeting the usual

© Springer International Publishing AG, part of Springer Nature 2018
G. Brazauskas et al. (Eds.): *Breeding Grasses and Protein Crops in the Era of Genomics*, pp. 14–19, 2018.
https://doi.org/10.1007/978-3-319-89578-9_3

requirements for agronomic performances, related notably to forage yield and feeding quality (Conaghan and Casler 2011).

Herbage production and feeding quality consistently depend on spring growth dynamic and earliness of heading (Dent and Aldrich 1963). A thorough assessment of these traits is therefore crucial to evaluate accessions of potential interest in breeding. We report hereafter field observations of spring growth and earliness of heading in a set of *L. perenne* genebank accessions that aimed at depicting the wide range of ecotypic variability existing in this species in Europe.

2   Material and Methods

2.1   Plant Material and Phenotyping

Fourteen European genebanks contributed to the set of *L. perenne* accessions under study, which included 419 natural populations and 14 cultivars. Micro-sward trials were established in three locations at Lusignan in France "LUS" (0.08 E; 46.40 N), Melle in Belgium "MEL" (3.79 E; 50.98 N) and Poel island in Germany "POE" (11.46 E; 54.00 N). The number of sown accessions was 427, 391 and 433 at LUS, MEL and POE, respectively. In each location, accessions were sown in three complete blocks. Elementary micro-plot size was 1 m². LUS and POE were sown in April 2015; MEL was sown in October 2015. Trials were fertilized with mineral nitrogen (200 to 300 kg N ha^{-1} split in several dressings in each location in 2016). In 2016, spring growth was measured weekly as sward canopy height using a herbometer. This consisted of a weighted plate on a graduated pole. The measurement thus integrates over the height and the density of the vegetation layer (Holmes 1974). Sward height was measured as undisturbed height at POE holding the herbometer plate at the top of the canopy, whereas it was measured by letting the plate settle on the canopy at LUS and MEL. Sward height measurement started at the onset of growth in all locations. It was performed weekly until heading date at LUS and POE and until the 3rd of May (before completion of heading) at MEL. Heading date was recorded at LUS and POE as the date when the ears of 20 plants were visible in a micro-plot. At each location, growing degree days (GDD) were computed starting on 01/01/2016 and taking 0 °C as the base temperature.

2.2   Data Analysis

We chose the Schnute's non-linear model (Schnute 1981) to fit spring growth data. Due to its flexibility, this model is able to fit various shapes of response curves as it was necessary with our data. It is expressed as:

$$Y_i = \left(c^b + (d^b - c^b)\left(\frac{1 - \exp^{-a(X-x1)}}{1 - \exp^{-a(x2-x1)}}\right)\right)^{\frac{1}{b}} + \in_i$$

where Y is the sward height; X is GDDs at time of measurement, and a and b are the parameters determining the shape of the curve. x1 and x2 are respectively the first and last values of the X domain for which the model is fitted; parameters c and d are the values of Y at x1 and x2 respectively. ε_i is the experimental error of the i measurement. Parameters a and b were estimated using the Levenberg–Marquardt iterative method with automatic computation of the analytical partial derivatives (SAS 8.01, SAS Institute, Cary, NC, USA). Maximum growth rate (Growth$_{max}$) was determined as when the second derivative of the Schnute's model function equaled zero. X and Y values at this point of the modelled curve gave the thermal time when the growth rate was maximum (GDD$_{growth.max}$) and the sward height at maximum growth rate (Height$_{growth.max}$), respectively. This model was implemented using a custom R script. In cases where the Schnute model took the shape of an exponential curve, the analytical result for GDD$_{growth.max}$ and Height$_{growth.max}$ became infinite. In these cases, for comparisons purposes, we considered GDD$_{growth.max}$ to be x2 and Height$_{growth.max}$ to be d. The Schnute model was used to predict growth rates for X = 200, 400, 600 and 800 GDD (Growth$_{200–800}$), the latter corresponding to the date of the latest measurements at MEL.

Genetic correlations between heading data in thermal time (HEA$_{GDD}$), Growth$_{max}$, Height$_{growth.max}$ and GDD$_{growth.max}$ were computed for LUS and POE separately. These correlations were obtained from a multivariate mixed ANOVA model in which the accession effect was considered as random. Phenotypic correlations were also computed in LUS and POE as pairwise Pearson correlations based on the means of accessions over the three replicates.

3 Results

Spring growth dynamics contrasted between trial locations. Figure 1 shows the distribution of accession growth rates in each trial site at different thermal times. The average growth rate (mean of accessions) continuously increased from 200 to 800 GDD at MEL and LUS (Fig. 1 A-D), whereas it showed a maximum round 500 GDD at POE (Fig. 1G). Figure 2G also shows that the average growth rate increased until 1000 GDD at LUS. Furthermore, the average maximum growth rate was found higher at MEL than at LUS (Fig. 1, E) but the average sward height at maximum growth rate than at LUS a smaller value at MEL than at LUS (Fig. 1, F).

Heading occurred later at LUS than at POE in terms of GDDs (Fig. 2) but it was the reverse in terms of calendar days (heading 5 days earlier on average in LUS). Genetic correlations were fairly comparable to phenotypic correlations (Table 1), indicating small residual variation in the phenotypic means of accessions per location. In LUS, the thermal time to heading of accessions (HEA$_{GDD}$) was strongly positively correlated to their thermal time to maximum growth rate (GDD$_{growth.max}$). Meanwhile, in this location, thermal times to heading and to maximum growth rate were negatively correlated to maximum growth rate (Growth$_{max}$). However, HEA$_{GDD}$ and GDD$_{growth.max}$ exhibited moderate positive correlations with sward height at time of maximum growth rate (Height$_{growth.max}$), because of the longer growing period of late accessions. In contrast, HEA$_{GDD}$ was negatively correlated to GDD$_{growth.max}$, in POE. In this location, Growth$_{max}$ was still negatively correlated to HEA$_{GDD}$ but it was positively correlated to GDD$_{growth.max}$. In POE, Height$_{growth.max}$ was finally positively correlated with GDD$_{growth.max}$ but not with HEA$_{GDD}$.

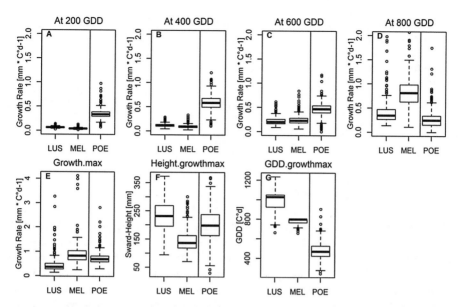

Fig. 1. A-D: Growth rates at the three locations MEL, LUS and POE from 200 to 800 growing degree-days (GDD); **E**: Maximum growth rates (Growth$_{max}$) at the three trial sites; **F**: Sward heights at time of maximum growth rate (Height$_{growth.max}$) at the three trial sites; **G**: Thermal time (growing degree-days) at time of maximum growth rate (GDD$_{growth.max}$) at the three trial sites. Figures show the distribution of accessions means over the three replicates per trial sites. Note that measurements at LUS and MEL are not formally comparable to those at POE due to different measurement techniques.

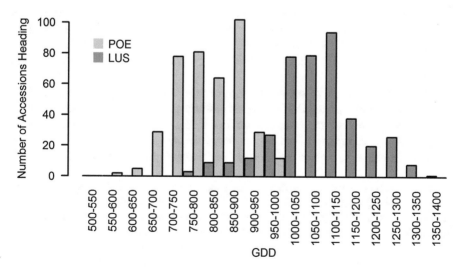

Fig. 2. Histogramme of heading time in growing-degree-days (GDD) at POE and LUS. Light grey bars refer to POE and dark grey bars to LUS

Table 1. Correlations between spring growth traits (Growth$_{max}$, Height$_{growth.max}$, GDD$_{growth.max}$) and thermal time to heading (HEA$_{GDD}$) in the trial sites POE and LUS. Genetic and phenotypic correlations are displayed below and above the diagonal of tables, respectively. The level of significance of phenotypic correlations is displayed as "*" for p value < 0.05 and "***" for p value < 0.001.

POE	GDD$_{growth.max}$	Height$_{growth.max}$	Growth$_{max}$	HEA$_{GDD}$	LUS	GDD$_{growth.max}$	Height$_{growth.max}$	Growth$_{max}$	HEA$_{GDD}$
GDD$_{growth.max}$	1	0.791***	0.123*	−0.184***	GDD$_{growth.max}$	1	0.486***	−0.624***	0.894***
Heigh$_{growth.max}$	0,637	1	0.428***	−0.316***	Height$_{growth.max}$	0,486	1	0.032	0.278***
Growth$_{max}$	0,243	0,732	1	−0.561***	Growth$_{max}$	−0,654	0,022	1	−0.687***
HEA$_{GDD}$	−0,323	−0,465	−0,845	1	HEA$_{GDD}$	0,926	0,302	−0,725	1

4 Discussion and Conclusion

This study presents preliminary results of spring growth and phenology for a wide range of *L. perenne* ecotypes. We cannot rule out that different modes of height measurements using the herbometer might have interfered with our results. However, it is clear that the dynamics of spring growth along thermal time was different in the three locations. It is, more specifically, a matter of fact that the genetic correlation between thermal times to heading and to maximum growth rate changed from strongly positive in LUS to moderately negative in POE. This is evidence of a strong interaction between accessions and trial sites in the timing of spring growth. The late heading accessions reached their maximum growth rate at smaller amount of degree-days in POE than in LUS, whereas early accessions did at similar amount of degree-days in the two trial sites (data not shown). This indicates that late accessions suffered from a limiting factor in POE which depressed their canopy growth quite soon until the end of the canopy height measurement period. We indeed used a very wide set of accessions covering almost the full range of heading earliness of perennial ryegrass. The late accessions we used were mostly ecotypes from oceanic climate (Western Europe) with warmer winter and spring than in POE. It is thus likely that the growth of late accessions was depressed in POE by insufficient delivery of growing-degree-days and/or by cold stresses.

The data used for the reported investigations was recorded within the frame of the project *GrassLandscape* which investigates climatic adaptation in the natural diversity of *L. perenne* at phenotypic and genomic levels. Next developments in the course of this project will contribute to better understand the response of spring growth to climatic conditions in *L. perenne* and the genetic variability of this response.

Acknowledgments. Results presented in this paper were obtained within the frame of the *GrassLandscape* project awarded by the 2014 FACCE-JPI ERA-NET + call *Climate Smart Agriculture*. (http://www.faccejpi.com/Research-Themes-and-Achievements/Climate-Change-Adaptation/ERA-NET-Plus-on-Climate-Smart-Agriculture/Grasslandscape)

References

Boller B, Greene SL (2010) Genetic resources. In: Boller B, Posselt UK, Veronesi F (eds) Fodder crops and amenity grasses. Springer, New York, pp 13–37

Conaghan P, Casler MD (2011) A theoretical and practical analysis of the optimum breeding system for perennial ryegrass. Ir J Agric Food Res. 50(1):47–63

Dent JW, Aldrich DTA (1963) The inter-relationships between heading date, yield, chemical composition and digestibility in varieties of perennial ryegrass, timothy, cocksfoot and meadow fescue. J. Nat Inst Agric Bot 9:261–281

Holmes CW (1974) The Massey grass meter. In: Dairy farming annual. Massey University, Palmerstone North, New Zealand, pp 26–30

Humphreys M, Feuerstein U, Vandewalle M, Baert J (2010) Ryegrasses. In: Boller B, Posselt UK, Veronesi F (eds) Fodder crops and amenity grasses. Springer, New York, pp 211–260

Sampoux JP, Baudouin P, Bayle B, Béguier V, Bourdon P, Chosson JF, Noël D (2013) Breeding perennial ryegrass (Lolium perenne L.) for turf usage: an assessment of genetic improvements in cultivars released in Europe, 1974–2004. Grass Forage Sci 68(1):33–48

Schnute J (1981) A versatile growth model with statistically stable parameters. Can J Fish Aquat Sci 38(9):1128–1140

Population Structure of Red Clover Ecotypes Collected from Europe and Asia

C. Jones[1(✉)], J. De Vega[2], D. Lloyd[1], M. Hegarty[1], S. Ayling[2], W. Powell[3], and L. Skøt[1]

[1] Institute of Biological, Environmental and Rural Sciences, Aberystwyth University, Gogerddan, Aberystwyth SY23 3EE, UK
{cgj,lfs}@aber.ac.uk
[2] The Earlham Institute, Norwich Research Park, Norwich NR4 7UH, UK
[3] Scotlands Rural College (SRUC), Peter Wilson Building, Kings Buildings, West Mains Road, Edinburgh EH9 3JG, UK

Abstract. Red clover (*Trifolium pratense* L.) is a highly adaptable species, with a wide geographic distribution. At the genetic level, wild ecotypes possess high variability, which enables them to adapt rapidly to diverse environments. This variation can be identified, via molecular techniques, and used to determine the differences between populations that may otherwise be indistinguishable. The results presented here are part of a wider study undertaken with a view to assess the available genetic diversity and use it to facilitate future introgression of useful traits into red clover breeding populations.

Keywords: *Trifolium pratense* · Ecotypes · Genotyping by sequencing
Allele frequency

1 Introduction

Cultivated red clover is used successfully as a silage crop, largely due to its tall growth habit and high yield (12t/ha/y). However, like many perennial forage species, red clover has limitations in its persistency. It will generally survive up to three years in the field. Added to this the central crown is unable to regenerate new growth after damage (Taylor and Quesenberry 1996), which limits red clover value as a grazed forage. This is a key trait for improvement, and could be achieved via the integration of a more prostrate habit, along with the ability to both branch and root at the nodes. Natural populations of red clover have varied growth habits and under certain conditions may produce nodal root growth (Frame et al. 1998).

It is expected that wild populations of red clover will show genetic variation that can be used to determine population structure and subsequently genetic diversity. The genetic diversity of wild and cultivated populations of red clover have previously been studied using low throughput methods (Ahsyee et al. 2014). Genotyping by sequencing (GBS) (Elshire et al. 2011) has become a popular high throughput technique to identify large scale variation across entire genomes. These whole genome scale SNP loci will greatly enhance the understanding of molecular diversity, and enable the dissection of complex traits, especially those involved in adaptive selection.

© Springer International Publishing AG, part of Springer Nature 2018
G. Brazauskas et al. (Eds.): *Breeding Grasses and Protein Crops in the Era of Genomics*, pp. 20–26, 2018.
https://doi.org/10.1007/978-3-319-89578-9_4

There is a large germplasm collection of wild red clover at IBERS, originating from various countries and distinctive habitats across the Middle East, Europe and Central and Western Asia. The collection has ecotypes with prostrate growth habit and nodal rooting, the divergence from which is a key domestication trait in red clover. The issue is how to make best use of the wealth of diversity in the germplasm available. This diversity can be used to identify changes in haplotype structure in response to biotic, eco-geographical, climatic and other abiotic factors shaping the adaptation of red clover to novel environments, and the effect of breeding and genetic improvement on allelic frequency.

The aim of this project was to develop a diverse ecotype panel from the collections of wild red clover held at IBERS along with elite breeding material and to use it to understand and characterize the genetic diversity within red clover, including differences in allele frequency. The data collected were used to compare the genetic diversity within and between the ecotypes and the breeding lines. This enabled us to detect population structure, chromosomal regions potentially under selection, and to identify SNP relating to phenotype. The work presented here is limited to the analysis to elucidate population structure and genomic regions under selection.

2 Materials and Methods

Ecotype Panel: Eight individuals from 70 ecotype accessions and 5 elite breeding lines were planted in the field in May 2015. The ecotype accessions came from Scandinavia, Europe, Iran, and Asia (Fig. 1).

Fig. 1. Country origin of each collection site used in the ecotype panel.

The five elite breeding lines are all of European descent (Table 1). Broadway and Crossway are early generation varieties developed at AgResearch (NZL) from ecotype collections from Portugal and Spain. These plants are low growing and under certain damp conditions produce stems that root (Rumball et al. 2003).

Table 1. The five varieties used in the ecotype panel. The table shows the place of varietal development and their origin

Variety	Development	Region
AberRuby	IBERS, GBR	NE
Britta	Uppsala, SWE	NE
Broadway	Ag-Research, NZL	SE
Crossway	Ag-Research, NZL	SE
Milvus	Agroscope, CHE	SCE

GBS Protocol and SNP Discovery: DNA was prepared for sequencing from 100 mg of fresh tissue following the published GBS protocol (Elshire et al. 2011), with modifications. TASSELv5 (Glaubitz et al. 2014) and BWA (Li and Durbin 2009) were used in conjunction with the red clover genome (De Vega et al. 2015) for a reference based GBS pipeline to identify high quality SNP within the ecotype panel. Table 2 shows the parameters used for the SNP calling.

Table 2. SNP calling parameters used in TASSELv5. The SNP had to pass all of these criteria to reach the final count and to be used in the genetic diversity analysis.

Parameter	Value
Tag length	20–64 bp
Quality score	20
Minimum reads per SNP site	10
Minor allele frequency	0.05
SNP present in genotypes	80% of genotypes
Missingness	0.1 per genotype

Population Structure: Was determined using allele frequencies by PCA analysis, STRUCTUREv2.3 and STRUCTURE harvester web. The population genetic parameters, using the allele frequencies, were calculated in R-packages Pegas, Hierfstat and StAMPP. To identify candidate loci potentially influenced by selection a three way comparison was made between loci identified using basic R scripts (Team 2014), BayeScan v2.0 (Foll and Gaggiotti 2008), and Samβada v0.5.3 (Stucki et al. 2016). R and BayeScan use the population parameter F_{ST} to identify loci under selection, Samβada uses a spatial correlation method.

3 Results and Discussion

The sequencing produced an average of 1,000,000 reads per sample. This was as expected, as one sample per genotype was sequenced. TASSELv5 identified around 1,8M tags across the whole population, with around 215,000 SNP within these tags. The SNP data were trimmed for minor allele frequency, reads per SNP, missingness and % genotypes covered (Table 2). This cleanup resulted in around 8,000 robust and reliable SNP with a position in the pseudomolecules of the red clover genome. These 8,000 SNP were then used in the genetic diversity analysis.

The PCA (Fig. 2) indicated a two-population structure to the data, split in an east – west orientation. The accessions on the left arm of the PCA were from Iran and Asia, and those on the right arm were of European descent. The allele frequencies calculated for each SNP in the accessions in Europe showed detectable variation between Mainland Europe (average 0.18) and those collectively from SWE, NE and the varieties Crossway and Broadway (average 0.20).

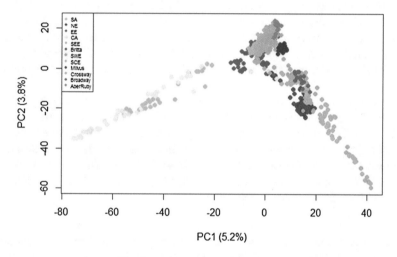

Fig. 2. PCA of the ecotype panel.

STRUCTURE harvester indicated a nine-population structure; however, there was a very large secondary peak at two populations. The two-population (k = 2) reflected the east – west split seen in the PCA. k = 9 subdivided the eastern accessions into two groups of Asia and Iran, and the European accessions into those from the Iberian Peninsula, and Northern Europe. The other accessions were more ambiguous, providing potential evidence of inter accession pollinations and mixing.

The accessions were analyzed for genetic diversity using the 8,000 SNP loci isolated from the pseudomolecules. The data presented in Table 3, shows that the observed and expected heterozygosity within the accessions was not significantly different ($H_S = 0.24$, $H_T = 0.26$); the F_{IS} value indicated random mating within the accessions, consistent with an outcrossing mode of reproduction. Overall the accessions were moderately differentiated ($F_{ST} = 0.0731$), indicating a population on the whole that was fairly related. This is reflected in the PCA by the clustering of SCE, SEE and varieties (Fig. 2), and in the ambiguous nature of the STRUCTURE harvester results. In that, the Mainland European accessions could not be reliably separated from each other.

Table 3. Basic statistics generated during genetic diversity analysis.

Statistic	Value
H_S	0.2392
H_T	0.2581
D_{ST}	0.0189
F_{ST}	0.0731
F_{IS}	−0.3408

The results in Table 3 show that there was relatively high heterozygosity seen within the accessions (H_S), but low heterozygosity between the accessions (D_{ST}); this is typical of old established natural populations, with high rates of gene flow which is probably a result of seed and pollen movement. However these levels of variation were high compared to other widely distributed ($H_T = 0.202$) and outcrossing species ($H_T = 0.167$) this reflected the wide geographic distribution and outcrossing nature of red clover.

The pairwise F_{ST} vales between the accessions was calculated using the R package StAMPP. This analysis produced an F_{ST} value of 0.56 between the accessions at the tips of the PCA clusters, indicating a high genetic diversity between these two accessions. These two accessions were also phenotypically diverse, and showed some of the extremes of the phenotypic range (Table 4). The F_{ST} values between the Asian and Iranian accessions to those in Europe ranged from 0.26 to 0.56, describing little mixing between the continental accessions.

Table 4. Phenotype ranges of four traits in two contrasting accessions. These two accessions fall at the tips of the PCA clusters, and show the highest genetic diversity in terms of pairwise F_{ST} at 0.56. CA-1 Central Asian accession, SWE-1 Southwestern European accession.

Measurement	CA-1	SWE-1
Height (cm)	33–53	12–33
Width (cm)	30–70	47–90
Stem number	9–31	39–100
Fresh weight (g)	121–323	99–270

Of interest to genetic diversity is the detection of outlier markers. These markers are an indication of regions potentially under selection. For this analysis, all of the samples were used with all 8,000 SNP markers. Using the F_{ST} values with a basic R script, a region of particular interest was identified in the Iranian accessions on chromosome seven that was above the significance threshold (Fig. 3).

The top of the peak is SNP position Chr7_5219303. This SNP is within gene number 38211, a 7 transmembrane MLO family protein which is involved in mildew resistance. There is evidence in the field to suggest that the Iranian accessions do have a degree of mildew resistance over other ecotype accessions. However, this has not been verified under experimental conditions.

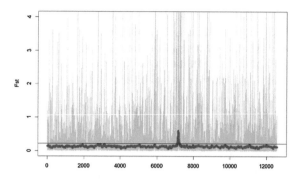

Fig. 3. The red line peak above the significance line (0.19) indicates outlier SNP, for Iranian accession Aa3507.

When the results from each of the three outlier detection methods (R, BayeScan, Samβada) were compared, 56 SNP loci were identified as being common to all three. Within this common group, loci were found in genes such as transcription factors involved in growth responses and flower development; genes involved in metabolic processes; and genes involved in various binding activities. However, these SNP are still putative loci that may or may not have been under selection pressure. The use of Samβada may have provided SNP that are under selection due to the geographic location of the accessions, whereas R and BayeScan provided evidence for variation in SNP allele frequencies as a means to detect outliers.

4 Conclusion

The centre of diversity for red clover is thought to be somewhere in the Eastern Mediterranean (Vavilov 2009), it is not known how it spread to its current distribution. This data would appear to support distinguishable populations in Europe and Asia, with little to no mixing after isolation. The data also support the proposal that Northern European and Iberian populations are more likely to be related to each other, than to those from the rest of Europe (Kjærgaard 2003). This is shown in the PCA (Fig. 2) in which Iberian and Northern European accessions are distinct from those from other parts of Europe, also consistent with the results of STRUCTURE harvester.

The accessions from Asia, Iran, and the Iberian Peninsula are isolated by their geography, at a global and local level. These accessions were collected from high mountain regions which has had an inevitable effect on population inter-pollination, as pollinating bees will travel a maximum of around 13 km from hive to flower source. This limits gene flow, and influences allele frequency change and stability. This may have led to population isolation and ultimately phenotypic adaptation to the environment.

Our results have indicated that there are potential SNP loci that are under diversifying selection because of local adaptation to the environment. The 56 common SNP identified have potentially aided the spread of red clover into the many diverse environments that it now occupies.

The loci putatively under selection may provide interesting sources of diversity for further study with a view to increase the effectiveness of red clover varieties. Added to this, it may be possible to use the ecotypes showing prostrate growth and nodal rooting. The integration of these traits, within the red clover breeding program at IBERS, into high yielding varieties may extend the use of this important forage legume.

Acknowledgements. This project is funded by BBSRC grants BB/L023563/1 to IBERS and BB/L022257/1 to Earlham, and by Germinal Holdings.

References

Ahsyee RS, Vasiljevic S, Calic I, Zoric M, Karagic D, Surlan-Momirovic G (2014) Genetic diversity in red clover (*Trifolium pratense* L.) using SSR markers. Genetika-Belgrade 46(3):949–961. https://doi.org/10.2298/gensr1403949a

De Vega JJ, Ayling S, Hegarty M, Kudrna D, Goicoechea JL, Ergon A, Rognli OA, Jones C, Swain M, Geurts R, Lang C, Mayer KFX, Rössner S, Yates S, Webb KJ, Donnison IS, Oldroyd GED, Wing RA, Caccamo M, Powell W, Abberton MT, Skøt L (2015) Red clover (*Trifolium pratense* L.) draft genome provides a platform for trait improvement. Sci Rep 5. https://doi.org/10.1038/srep17394

Elshire RJ, Glaubitz JC, Sun Q, Poland JA, Kawamoto K, Buckler ES, Mitchell SE (2011) A robust, simple Genotyping-by-Sequencing (GBS) approach for high diversity species. PLoS ONE 6(5). https://doi.org/10.1371/journal.pone.0019379

Foll M, Gaggiotti OE (2008) A genome scan method to identify selected loci appropriate for both dominant and codominant markers: a Bayesian perspective. Genetics 180:977–993

Frame J, Charlton JFL, Laidlaw AS (1998) Temperate forage legumes. CAB International, Wallingford

Glaubitz JC, Casstevens TM, Lu F, Harriman J, Elshire RJ, Sun Q, Buckler ES (2014) TASSEL-GBS: a high capacity genotyping by sequencing analysis pipeline. PLoS ONE 9(2). https://doi.org/10.1371/journal.pone.0090346

Kjærgaard T (2003) A plant that changed the world: the rise and fall of clover 1000–2000. Landsc Res 28:41–49

Li H, Durbin R (2009) Fast and accurate short read alignment with Burrows-Wheeler transform. Bioinformatics 25(14):1754–1760. https://doi.org/10.1093/bioinformatics/btp324

Rumball W, Keogh RG, Miller JE (2003) 'Crossway' and 'Grasslands Broadway' red clovers (Trifolium pratense L.). N Z J Agric Res 46(1):57–59

Stucki S, Orozco-terWengel P, Forester BR, Duruz S, Colli L, Masembe C, Negrini R, Landguth E, Jones MR, Bruford MW, Taberlet P, Joost S (2016) High performance computation of landscape genomic models including local indicators of spatial association. Mol Ecol Resourc. https://doi.org/10.1111/1755-0998.12629

Taylor NL, Quesenberry KH (1996) Red clover science. Kluwer Academic Publishers, Dordrecht

RCD Team (2014) R: a language and environment for statistical computing, Vienna, Austria

Vavilov NI (2009) On the eastern centers of cultivated plants. Cambridge University Press, Cambridge

Enhancing the Sustainable Use of *Lolium perenne* Genetic Resources from Genebanks in Plant Breeding and Research

S. Bachmann-Pfabe[1], E. Willner[2(✉)], M. Oppermann[3], S. Weise[3], and K. J. Dehmer[1]

[1] Leibniz Institute of Plant Genetics and Crop Plant Research (IPK) Gatersleben, Genebank, Satellite Collections North, Parkweg 3a, 180190 Groß Lüsewitz, Germany
[2] Leibniz Institute of Plant Genetics and Crop Plant Research (IPK) Gatersleben, Genebank, Satellite Collections North, Inselstraße 9, 23999 Malchow (Poel), Germany
willner@ipk-gatersleben.de
[3] Leibniz Institute of Plant Genetics and Crop Plant Research (IPK) Gatersleben, Genebank Documentation, Corrensstraße 3, 06466 Stadt Seeland, Germany

Abstract. *Lolium perenne* is one of the most important grass species in Europe and subject of intensive breeding and research. Hence, there is a substantial need for well described, diverse and easily available genetic resources. The Satellite Collections North of the IPK Genebank continuously characterize and evaluate genebank accessions of oil and fodder crops. As an example, results from characterization and evaluation trials of 336 *L. perenne* accessions collected in Bulgaria, Croatia, Ireland and Spain are presented. Traits considered were, amongst others, phenological development, ear emergence, biomass production, disease susceptibility and winter hardiness. Results from the three-year evaluation trials showed a high variability regarding these traits. Generally, the Spanish ecotypes revealed a high biomass production level, while the Bulgarian ecotypes were characterized by a low rust susceptibility. The Irish ecotypes showed a better growth after winter than the ecotypes from the other countries in the third scoring year. These results have been transferred to the in-house developed Genebank Information System (GBIS) and to the international EURISCO information system. Their access via EURISCO is demonstrated.

Keywords: Perennial ryegrass · Phenotypic traits · Variability
Data management · GBIS · EURISCO

1 Introduction

Perennial ryegrass (*Lolium perenne* L.) is one of the most important species for grasslands in temperate regions all over the world. Its high feeding value makes it a valuable forage crop for ruminants, but it is also well suited for turf usage. Nowadays, ryegrass breeding focus on traits such as persistence, winter hardiness and resistance to abiotic and biotic stress (Sampoux et al. 2011; Sanna et al. 2014). Sources for new

© Springer International Publishing AG, part of Springer Nature 2018
G. Brazauskas et al. (Eds.): *Breeding Grasses and Protein Crops in the Era of Genomics*, pp. 27–32, 2018.
https://doi.org/10.1007/978-3-319-89578-9_5

germplasm with the desired properties might be genebank collections of *L. perenne*. The genebank collection of the Leibniz Institute of Plant Genetics and Crop Plant Research (IPK) is the largest of its kind in Western Europe totalling over 150,000 accessions. It comprises a comprehensive grass collection, and amongst them more than 3,000 *L. perenne* accessions are currently being maintained. The conservation work focusses on keeping a large majority of accessions available as high quality seed stocks (91%), recording basic characteristics of the accessions as well as results from evaluation trials, and making these information available for interested users, such as scientists and plant breeders. The aim of this paper is (I) to describe the phenotypic variability in *L. perenne* genebank accessions collected in different European countries and (II) to show how these results can be easily accessed via international information systems such as EURISCO in order to facilitate and enhance the utilization of these resources.

2 Material and Methods

The performance of *L. perenne* ecotypes originating from collection trips to Bulgaria, Croatia, Ireland and Spain was studied in characterization and evaluation trials at the Malchow/Poel site of the IPK in Northern Germany[1]. *Characterization:* The collected ecotypes were cultivated in isolated plots of 4×6 meters size in the field for regeneration. As fertilizer nitrogen (N) was applied in the form of calcium ammonium nitrate at the beginning of the vegetation period (March, 80 kg N ha^{-1}) and after cuttings (60 kg N ha^{-1}). Phenological traits were recorded such as heading date (HD), flowering time (FT), begin of harvest (Harv B) and end of harvest (Harv E). *Evaluation:* The ecotypes were evaluated in four single field trials in comparison to standard varieties (Barclay, Fennema, Juwel, Sambin, Gladio, Respect, Cancan and Aberelf). The trials were arranged in a completely randomized block design and ran three years each (ecotypes from Bulgaria 2002–04, Croatia 2003–05, Spain 2004–06, Ireland 2008–10). Fertilizers were applied as described above. The mean daily air temperature during the experimental period ranged between 8.5 and 10.7 °C, with mild winters and hot summers in 2008 and 2009. The average annual rainfall ranged between 462 and 816 mm and was low in the year 2008 and high in 2002, particularly from July to October. Agronomic traits were scored on a scale ranging from 1 (low) to 9 (high): flowering tendency in the year of sowing (FLW1), growth before winter in the year of sowing (GBW1) and the second experimental year (GBW2), growing in spring in the second (GIS2) and third experimental year (GIS3), aftermath heading (AH2), aboveground plant biomass (PBM2), rust and disease susceptibility (RUST2, DIS2) in the second experimental year and growth after winter in the third experimental year (GAW3). To allow for a rough comparison between the four trials, results were corrected for the "year effect" based on the standard varieties as described in Bothe et al. (2016). Subsequently, the raw data collected in the course of these experiments were imported into the Genebank Information System (GBIS) of the IPK (Oppermann et al.

---

[1] Satellite Collections North, Longitude 11°28′26″ E, Latitude 53°59′40″ N, 10 m ASL, annual precipitation 530 mm, mean annual temperature 9.1 °C, soil type: sandy loam.

2015). This system has been developed to support all the complex processes necessary to maintain the IPK Genebank collections and to enable interested users to search for suitable germplasm and to order samples online. Moreover, GBIS is an important source for global aggregators, such as the Global Biodiversity Information Facility (GBIF) or the European Search Catalogue for Plant Genetic Resources (EURISCO).

3 Results and Discussion

The results of the trials revealed a high variability of IPK's *L. perenne* collection regarding several agronomic traits. As an example, the phenological development of the ecotypes collected in Bulgaria and Spain is presented in Table 1.

Table 1. Phenology of *L. perenne* ecotypes collected in Bulgaria and Spain as a result of the characterization trials at the Malchow (Poel) site. Results are presented in days after 1st of April.

Trait	Ecotypes collected in BGR (N = 60)			Ecotypes collected in ESP (N = 72)		
	mean	min/max	cv %	mean	min/max	cv %
HD 50	46.9	28/58	14	52.1	44/61	7.9
FT 50	69.1	58/78	6.7	77.9	67/94	8.4
Harv B	105	88/118	7.0	111	91/122	7.3
Harv E	112	89/123	6.1	124	102/148	8.4

HD – heading date 50%, *FT* – flowering time 50%,
Harv B – harvest date, beginning, *Harv E* – harvest date,
end; *BGR* = Bulgaria (characterized 1999–2000),
ESP = Spain (characterized 2000–2001), *cv %* - coefficient
of variation.

The Bulgarian ecotypes were characterized by a slightly earlier heading and a faster flowering in comparison to the Spanish ecotypes. The largest difference was observed in the period between begin and end of harvest, which took, on average, seven days for the Bulgarian ecotypes and 13 days for the Spanish ecotypes. This could be attributed to the collection sites of the Bulgarian ecotypes which mainly originate from mountainous regions or flood plains. Nevertheless an influence of different climate conditions during the regeneration trails cannot be clearly excluded yet.

In the evaluation trials, a high variation was observed for the traits FLW1, AH2, RUST2, DIS2 and GAW3 (Table 2). The biomass production level (GBW, GIS, PBM) of the ecotypes was comparable to that of the standard varieties. In particular the Spanish ecotypes revealed a high biomass production throughout the experimental period. The ecotypes collected in Bulgaria were characterized by a low plant biomass production (PBM2) and a low susceptibility to rust infection (RUST2) in the second scoring year. In regard to the growth after winter (GAW3), the ecotypes partly exceeded the performance of the standard varieties, what can mainly be attributed to the superior growth of the Irish ecotypes in the third scoring year.

Table 2. Variation of phenotypic traits among 336 perennial ryegrass (*L. perenne*) ecotypes collected in Bulgaria, Spain, Croatia and Ireland evaluated in three-year field trials in comparison to commercial cultivars. (Bothe et al. 2016, adapted)

Year	Trait	Ecotypes (N = 336)			Cultivars (N = 32)		
		mean	min/max	cv %	mean	min/max	cv %
1	FLW	1.2	0.6/5.9	44	1.1	0.6/1.8	20
1	GBW	7.3	5.3/9.1	9.6	7.8	5.8/9.6	9.2
2	GIS	6.9	3.6/9.6	15	7.0	4.4/8.2	14
2	PBM	5.1	1.5/6.7	19	5.8	3.7/7.7	16
2	AH	2.6	0.5/5.4	45	2.8	2.2/4.2	16
2	RUST	3.1	1.4/4.9	27	3.6	2.6/5.2	12
2	DIS	3.3	2.1/5.6	20	3.2	2.6/4.0	8.5
2	GBW	6.5	3.2/8.2	10	6.7	5.6/7.8	8.5
3	GAW	4.6	2.5/7.2	22	4.4	2.7/5.5	13
3	GIS	5.5	2.5/7.9	18	5.9	3.2/7.0	17

FLW – flowering tendency, *GBW* – growth before winter, *GIS* – growing in spring, *PBM* – plant biomass, *AH* – aftermath heading, *RUST* – rust susceptibility, *DIS* – disease susceptibility, *GAW* – growth after winter, *cv %* – coefficient of variation.

In order to share the detailed data collected, it was first documented in IPK's GBIS database and then provided to the European Search Catalogue for Plant Genetic Resources (EURISCO; Weise et al. 2017). EURISCO is an information system providing information of about 1.9 million accessions of crop plants and their wild relatives, preserved *ex situ* by almost 400 institutions. It is maintained by the IPK Genebank on behalf of the European Cooperative Programme for Plant Genetic Resources (ECPGR) and bases on a network of *ex situ* National Inventories of 43 member countries.

Presently, the accessions documented in EURISCO comprise more than 6,200 genera. The systems contains both passport, characterization and evaluation data about plant genetic resources, which is being updated regularly. The public web interface of EURISCO enables the users to filter the available phenotypic data by means of four wizards. The search can be started by selecting (1) experiment, (2) trait, (3) genus as well as (4) combinations of species and traits. In addition, user-specific filter rules can be defined. Besides the observed phenotypic scores, some descriptive statistics are generated for the selected traits (Fig. 1).

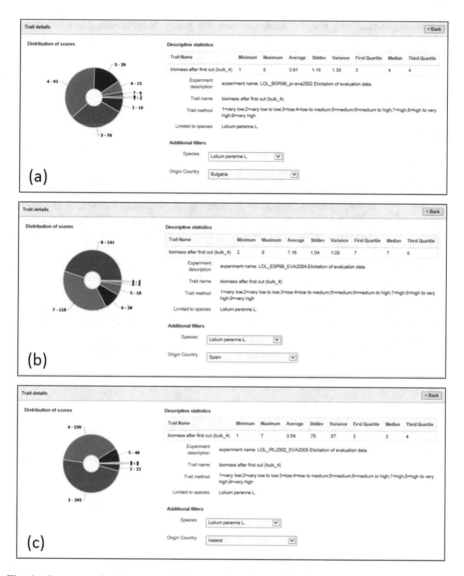

Fig. 1. Summary of rating scores of biomass after first cut showing *L. perenne* accessions from Bulgaria (a), Spain (b) and Ireland (c), respectively. The numbers of observations include replications.

4   Conclusion

In conclusion, the *L. perenne* collection of the IPK Genebank is well described and shows a high variability of important agronomic traits. According data can be easily accessed via the Genebank Information System (GBIS) and via the international information system EURISCO.

Acknowledgements. We like to thank the research group Bioinformatics and Information Technology of the IPK Gatersleben for providing and maintaining the institute's database environment.

References

Bothe A et al (2016) Phenotyping genetic diversity of perennial ryegrass ecotypes (*Lolium perenne* L.). In: Roldán-Ruiz I, Beart J, Reheul D (eds) Breeding in a world of scarcity. Springer, Cham, pp 21–27

Oppermann M, Weise S, Dittmann C, Knüpffer H (2015) GBIS – the information system of the German Genebank. Database (Oxford), 2015:bav021

Sampoux JP, Baudouin P, Bayle B, Béguier V, Bourdon P, Chosson JF et al (2011) Breeding perennial ryegrasses for forage usage: an experimental assessment of trait changes in diploid perennial ryegrass (*Lolium perenne* L.) cultivars released in the last four decades. Field Crops Res 123(2):117–129

Sanna F, Franca A, Porqueddu C, Piluzza G, Re GA, Sulas L, Bullitta S (2014) Characterization of native perennial ryegrasses for persistence in mediterranean rainfed conditions. Span J Agric Res 12(4):1110–1123

Weise S, Oppermann M, Maggioni L, van Hintum T, Knüpffer H (2017) EURISCO: the European search catalogue for plant genetic resources. Nucleic Acids Res 45(D1): D1003–D1008

Tetraploid *Festuca apennina* is Prone to Produce Triploid Hybrid Progeny When Crossed with Diploid *Festuca pratensis*

B. Boller[1(✉)], T. Felder[1], and D. Kopecký[2]

[1] Agroscope, Reckenholzstrasse 191, 8046 Zürich, Switzerland
beat.boller@agroscope.admin.ch
[2] Institute of Experimental Botany, Centre of Plant Structural and Functional
Genomics, Šlechtitelů 31, 783 71 Olomouc-Holice, Czech Republic

Abstract. *Festuca apennina* De Not. is a species growing at altitudes of 1100 to 2000 m a.s.l. in different regions of Europe. To test cross fertility of different cytotypes of *F. apennina*, pairwise crosses were made with tetraploid and triploid *F. apennina*, as well as diploid and colchicine-induced tetraploid *Festuca pratensis* Huds. Tetraploid *F. apennina* appeared to be quite cross-fertile with diploid *F. pratensis* and produced triploid progeny in both cross directions. When *F. apennina* plants were pollinated with diploid *F. pratensis*, they yielded even more seed per inflorescence than those pollinated with another *F. apennina* plant. About 25% of the progeny of this cross were triploids; the remaining 75% were tetraploid, presumably resulting from self-pollination. Much less seeds were obtained on diploid *F. pratensis* pollinated with *F. apennina*, but the progeny consisted of as much (52%) triploid hybrids as of diploid selfs (48%). Seeds harvested on *F. apennina* giving rise to triploid progeny appeared to have no cold requirement for germination, as opposed to tetraploid progeny from self-pollination or crosses among *F. apennina*. Colchicine-induced tetraploids and diploids of *F. pratensis* were much less cross-fertile. Few viable seeds were obtained, with a great majority (91%) resulting from self-pollination, with only 9% triploid hybrids. The surprisingly high cross fertility of tetraploid *F. apennina* with diploid *F. pratensis*, the apparent absence of a triploid block in their progeny, and the fact that seeds of *F. apennina* giving rise to triploid progeny have no cold requirement, probably contribute to the frequent occurrence of triploid forms at the low end of *F. apennina* altitudinal distribution.

Keywords: Cross fertility · Diploid · *Festuca apennina* · *Festuca pratensis*
Ploidy · Tetraploid · Triploid

1 Introduction

Festuca apennina De Not. was originally described as a plant of shaded woodland-cleared habitats in the Apennine mountain range. However, very similar, and probably the same species was later found in nutrient rich zones of grassland in the Alps where it tended to dominate large areas. Ecologically and morphologically, plants used in this study correspond with this latter *Festuca pratensis* var. *megalostachys*

© Springer International Publishing AG, part of Springer Nature 2018
G. Brazauskas et al. (Eds.): *Breeding Grasses and Protein Crops in the Era of Genomics*, pp. 33–38, 2018.
https://doi.org/10.1007/978-3-319-89578-9_6

Stebler. They resemble *F. pratensis* Huds. but have wider leaves, thicker stems, and larger, slightly awned florets. Our previous study (Kopecký et al. 2016) has shown that the Apennine, Carpathian and Alpine accessions have the same genomic constitution. Thus, *Festuca apennina* De Not., *Festuca pratensis* var. *megalostachys* Stebler, and *Festuca pratensis* ssp. *apennina* (De Not.) Hegi, are very likely synonymous and from here on will be referred to as *F. apennina*. Kopecký et al. (2016) provided evidence that *F. apennina* is an allotetraploid hybrid between diploid *F. pratensis* and a hitherto unknown diploid species. Besides tetraploids, triploid *F. apennina*-like plants were found in nature by Tyler (1988). He hypothesized those were hybrids between tetraploid *F. apennina,* found at higher altitudes, and diploid *F. pratensis* in mid-altitudinal zones where both species co-exist sympatrically. Kopecký et al. (2016) provided GISH-based evidence that natural triploids are indeed a consequence of natural hybridization between *F. apennina* and *F. pratensis.*

The frequent occurrence of triploid *Festuca* in many Swiss locations studied by Kopecký et al. (2016), often in apparent absence of one or even both progenitors, is intriguing. It suggests that plants of different ploidy within the *Festuca* complex can be highly cross-fertile. The question rises whether triploid progenies originate from pollination of a tetraploid *F. apennina* with pollen from a diploid *F. pratensis*, vice-versa, or both, and which cross direction is more likely to produce hybrids. Finally, the fact that triploids can dominate swards in alpine pastures made us wonder if natural triploid *Festuca* are indeed as sterile as expected based on experience with synthetic triploids (Clarke et al. 1976).

To study this issue we performed controlled hybridizations involving diploid *F. pratensis*, colchicine-induced tetraploid *F. pratensis,* triploid *F. apennina* and tetraploid *F. apennina*. Seed yields and germination rates were determined, and progeny was raised to discriminate between hybrids and progenies generated by self-pollination.

2   Material and Methods

2.1   Plant material

Festuca apennina originated from the 2015 collection at Mörlialp, Glaubenbielen, Fontanen, Stoos, Moléson, and La Para (Kopecký et al. 2016). Diploid and colchicine-induced tetraploid *F. pratensis* plants were chosen from the Agroscope *Festuca* breeding programs. Diploid plants were derived from the ecotype "Pfisterboden", crossed and backcrossed with breeding material similar to cv. Preval. Tetraploids were of the C_2 generation (2 cycles of recurrent selection after colchicine treatment) tracing back to diploid breeding material selected for high molecular diversity.

2.2   Cross pollination

Pairwise crosses among selected 4x and 3x *F. apennina*, as well as 2x and colchicine-induced 4x *F. pratensis* genotypes were carried out with detached tillers, sampled before anthesis at full inflorescence emergence. For each possible combination of the four cytotypes, 4 to 5 pairs involving 4 to 5 different genotypes from each side were formed. With few exceptions, the same 5 genotypes of each cytotype class were used for the different combinations. No emasculation was done, but care was taken to make sure that no florets were already flowering, and that inflorescences would start flowering within 1 or 2 days after bagging to ensure flowering synchrony between the partners. However, pollen quantity and quality was not assessed and might have influenced the fertilization success of the male partners of the crosses. For each pair cross, between 1 and 3 tillers of each of the two genotypes were placed in reagent tubes filled with tap water and the inflorescences were enclosed in pollination bags. Bagged tillers were maintained in a Conviron growth chamber at 20/15°C day/night temperature and 70/80% relative humidity, and 450 μmol m^{-2} s^{-1} PAR during a 16 h photoperiod. 14 d after bagging, inflorescences of the two crossing partners were separated and enclosed individually in greaseproof bags to avoid mixing of prematurely shed seeds. When seeds were ripe, the inflorescences were dried and the seeds cleaned and counted. Seed yield was determined as number of seeds per inflorescence. Therefore, when comparing seed yield of mother plants of different ploidy, differences in the number of florets per inflorescence between the ploidy levels could have biased the results. However, the effect of the cytotype of the male partner should not be biased.

2.3   Germination trials and ploidy determination of progeny

Germination rate was determined on as many seeds as were available, with a maximum of 50 seeds. For comparison of germination rates, only mother plants having yielded at least 5 seeds in one particular cross were included. On average, 33 seeds per individual were used. Seeds were placed on wet filter paper in Petri dishes, in a first trial (referred to as "without pre-cooling"), maintained at room temperature. Germinated seedlings were planted in boxes filled with potting compost, as needed. After 14 d, remaining seeds were counted and the germination rate expressed as a percentage of germinated seeds in the total number of seeds. In the second test, remnant seeds of the same crosses, harvested from *F. apennina* plants, were imbibed and pre-cooled at 4°C for 14 d, before transferring to a germination cabinet at 25°C for another 14 d. The number of viable seeds per inflorescence was calculated for each mother plant individually from the total number of seeds and the germination rate determined on a subsample. Seedlings were grown in a greenhouse with supplemental light under a 16 h photoperiod.

Ploidy of up to 16 progeny plants (6.4 plants on average) per mother plant was determined by flow cytometry (for details see Kopecký et al. 2016). Two representative plants of each cytotype were checked by chromosome squash counting.

3   Results and Discussion

Surprisingly, the highest seed yield (number of seeds per inflorescence) of all cytotype combinations was obtained for tetraploid *F. apennina* pollinated with diploid *F. pratensis* (Table 1). This combination yielded a similar number of seed to diploid diploid cross in *F. pratensis* and significantly more than cross within tetraploid *F. apennina*. This was not true for the reciprocal cross. Diploid *F. pratensis* pollinated with *F. apennina* produced the least seed of all combinations. Ploidy of all progenies from crosses with triploids as male was the same as that of the respective mother plant (detailed data not shown). Therefore we assume that triploids were male sterile and the harvested seed resulted from self-pollination, thus seed set represents self-pollination potential of the females. This assumption should be validated by enforced selfings. However, later observations showed that anthers of triploids were sterile, thus no pollen of triploids should have influenced the assessment of the selfing potential of the partners. Accepting this, diploid *F. pratensis* had the strongest self-incompatibility and thus lowest self-fertility, while *F. apennina* appeared to be quite self-fertile in the absence of compatible foreign pollen. Colchicine induced tetraploid *F. pratensis* also appeared to be markedly more self-fertile than diploid *F. pratensis*. The cross of *F. apennina* × tetraploid *F. pratensis* seemed to be less compatible than a comparable cross with diploid *F. pratensis*. Indeed, in both cross directions, fewer of the harvested seed resulted from cross- than from self-pollination ("cross" with male sterile triploids). Therefore, when crossing *F. apennina* with tetraploid *F. pratensis*, the progeny should be tested for their hybridity. A single seed was harvested from a total of 76 inflorescences of triploid plants used for crosses in all combinations (i.e. 0.01 seed per tiller). This seed produced a plant which died before reaching the 5 leaf stage. Therefore, naturally occurring triploid *F. apennina* × *F. pratensis* can be considered largely sterile, both male and female.

The results of germination rate (Table 1) reflected the known cold pre-treatment requirement of seed harvested on *F. apennina* (Tyler et al. 1978). However, when

Table 1. Seed yield per inflorescence and germination ability of seeds harvested on 2x and 4x *Festuca pratensis/apennina* individuals when paired with a single 2x, 3x or 4x partner plant

Cytotype of mother plant	Cytotype of male partner			
	2x *F. pratensis*	3x *F. apennina*	4x *F. apennina*	4x induced *F.p*
Seed yield (number of seeds per inflorescence)				
2x *F. pratensis*	**74.5 ab**[a]	5.8 d	3.6 d	25.9 cd
4x *F. apennina*	100.5 a	55.6 bc	**51.3 bc**	20.5 cd
4x induced *F.p.*	16.5 cd	28.3 cd	25.4 cd	**49.2 bc**
Germination rate % without pre-cooling (after 14 d pre-cooling at 4°C)				
2x *F. pratensis*	**92.5 ab**	98.4 a	89.6 ab	10.7 f
4x *F. apennina*	31.7 e (76.1 abcd)	0 f (59.0 cd)	**1.0 f (69.0 bcd)**	5.6 f (54.5 d)
4x induced *F.p*	80.7 abc	96.2 a	96.2 a	**91.8 ab**
Viable seeds per inflorescence without pre-cooling (after pre-cooling)				
2x *F. pratensis*	**69.0 ab**	3.7 de	3.3 de	3.0 de
4x *F. apennina*	35.9 bcde (83.7 a)	0 e (45.9 bc)	**0.3 e (39.1 bcd)**	1.7 e (17.8 cde)
4x induced *F.p*	13.6 cde	27.7 cde	19.4 cde	**46.2 bc**

[a]Means of 4 to 5 mother plants in 4 to 5 pair crosses. Values not followed by the same letter are significantly ($p < 0.05$) different (Duncan's test). **Bold** figures indicate pollination with an individual of the same cytotype

pollinated with diploid *F. pratensis*, the cold requirement of seed of *F. apennina* was significantly reduced. Indeed, those were the only *F. apennina* seeds whose germination rate (32%) was significantly different from 0 without cold pre-treatment.

The second striking feature of these results was very poor germination of seeds harvested from diploid *F. pratensis* pollinated with tetraploid *F. pratensis*, reflecting a strong triploid block in this cross direction. The germination rate of seed from 4x induced *F. pratensis* following pollination with a diploid, was slightly, but not significantly reduced compared to that from a 4x induced x 4x induced cross.

A large proportion of the progeny of *F. apennina* × diploid *F. pratensis* crosses were true triploid hybrids (Table 2). Indeed, 52% of the little progeny of diploid *F. pratensis*, when pollinated with *F. apennina*, were triploids, and it was 83% in the reciprocal cross when progeny was obtained without pre-cooling. When pre-cooling was applied, only 25% of the progeny of seeds harvested on *F. apennina* after crossing with diploid *F. pratensis* were triploids. This suggests that the seed of *F. apennina* giving rise to 3x plants has no cold requirement for germination. The number of viable triploid seeds per inflorescence was estimated to be about 25 (29.7 based on the results without and 20.9 with pre-cooling) for the *F. apennina* parent, compared to only 2 for the diploid *F. pratensis* parent. When colchicine-induced tetraploid and diploid *F. pratensis* were crossed, 25% and 8% of the progeny were triploid for the 2x and 4x parent, respectively, and only between 0.6 and 2.4 viable seeds per inflorescence were estimated.

Table 2. Occurrence of triploid (3x) progeny from 2x *Festuca pratensis* × 4x *F. apennina/pratensis* crosses

Mother plant	Male partner	Cold treated	Number of progeny plants identified (viable seeds per inflorescence[a])		
			2x	3x	4x
2x *F. pratensis* × 4x *F. apennina* (total number of progeny from 5 pair crosses)					
2x *F. pratensis*	4x *F. apennina*	no	10 (1.6)	11 (1.7)	
4x *F. apennina*	2x *F. pratensis*	no		19 (29.7)	4 (6.2)
4x *F. apennina*	2x *F. pratensis*	yes		12 (20.9)	36 (62.8)
2x *F.pratensis* × 4x induced *F. pratensis* (total number of progeny from 5 pair crosses)					
2x *F. pratensis*	4x induced *F.p.*	no	4 (2.4)	1 (0.6)	
4x induced *F.p.*	2x *F. pratensis*	no		4 (1.2)	44 (12.5)

[a]calculated based on average number of viable seeds presented in Table 1

4   Conclusion

Despite the ploidy mismatch, tetraploid *F. apennina* hybridizes easily with diploid *F. pratensis* in both directions, but more viable triploid seed per inflorescence were obtained when *F. apennina* served as female. Seeds of *F. apennina* giving rise to 3x progeny have no cold requirement for germination, unlike 4x *F. apennina* progeny. Such behavior might facilitate the establishment of the triploid cytotype at the lower end of *F. apennina* altitudinal distribution.

References

Clarke J, Chandrasekharan P, Thomas H (1976) Studies in *Festuca* 9. cytological studies of *Festuca pratensis* var. *apennina* (De Not.) Hack. (2n = 28). Z Pflanzenzüchtg 77:205–214

Kopecký D, Harper J, Bartoš J, Gasior D, Vrána J, Hřibová E, Boller B, Ardenghi NMG, Šimoníková D, Doležel J, Humphreys MW (2016) An increasing need for productive and stress resilient Festulolium amphiploids: what can be learnt from the stable genomic composition of Festuca pratensis subsp. apennina (De Not.) Hegi? Front Environ Sci 4(66): 1–14

Tyler BF (1988) Description and distribution of natural variation in forage grasses. In: Natural Variation and Breeding for Adaptation, Proceedings of the Eucarpia Fodder Crops Section Meeting, 22–24 September 1987, Lusignan (France), p 13–22 (1988)

Tyler BF, Borrill M, Chorlton K (1978) Studies in Festuca. X. Observations on germination and seedling cold tolerance in diploid festuca pratensis and tetraploid F. pratensis var. apennina in relation to their altitudinal distribution. J Appl Ecol 15:219–226

Comparing Mitotic and Meiotic Tetraploid Red Clover Plants for Seed Yield

T. Vleugels[✉], I. Roldán-Ruiz, and G. Cnops

Plant Sciences Unit, ILVO (Flanders Research Institute for Agriculture, Fisheries and Food), Caritasstraat 39, 9090 Melle, Belgium
tim.vleugels@ilvo.vlaanderen.be

Abstract. New tetraploids can be created by chromosome doubling (mitotic tetraploids), or through unreduced gametes (meiotic tetraploids). In this study, meiotic tetraploids were generated in interploidy crosses, their seed yield was compared against mitotic tetraploids with similar genetic background, and the relation between genetic diversity and seed yield was explored. A total of 15 meiotic tetraploids were obtained in 16 interploidy crosses between diploid Lemmon or Merian genotypes and tetraploid Elanus genotypes. Mitotic tetraploids were created by crossing plants from tetraploidized Lemmon or Merian populations with Elanus genotypes. Tetraploid progeny from cross-pollinations were identified with flow cytometry and SSR marker analysis. The efficiency of interploidy crosses to generate meiotic tetraploids was low (0.5% in diploid and 0.7% in tetraploid mother plants) and the degree of self-pollination was high (95.0% in diploid and 94.1% in tetraploid mother plants). Using seeds from the diploid parent was more time and cost efficient because most undesired progeny was identified by flow cytometry, leaving fewer plants to be screened with SSR markers. Meiotic tetraploids yielded significantly more seeds than mitotic tetraploids with similar genetic background (15.9 g vs. 10.4 g), but their genetic diversity was not higher (2.6 vs. 2.9 alleles per SSR locus). Seed yield and the average allele number per locus were not correlated among all progeny plants.

Keywords: *Trifolium pratense* · Seed yield · Tetraploids · SSR markers
Unreduced gametes · Flow cytometry (FC)

1 Introduction

Numerous diploid and tetraploid red clover (*Trifolium pratense* L.) cultivars are available with excellent agronomic performance (Boller et al. 2010). While tetraploid cultivars generally outperform diploid ones for agricultural traits, their seed yield is often substantially lower. Diploid cultivars typically produce 400 to 500 kg seed/ha (Boller et al. 2010), while in tetraploid cultivars seed yield is 25 to 40% lower (Amdahl et al. 2015). As a consequence, seed production of tetraploid cultivars is more expensive and seed prices are higher (Taylor and Quesenberry 1996).

Tetraploidization of diploid germplasm can be obtained in two ways: through chromosome doubling using chemical agents such as colchicine (mitotic tetraploids), or through unreduced gametes in interploidy crosses (meiotic tetraploids) (Taylor and

© Springer International Publishing AG, part of Springer Nature 2018
G. Brazauskas et al. (Eds.): *Breeding Grasses and Protein Crops in the Era of Genomics*, pp. 39–45, 2018.
https://doi.org/10.1007/978-3-319-89578-9_7

Quesenberry 1996). It is assumed that mitotic tetraploids are less heterozygous than meiotic tetraploids because in the latter, part of the heterozygosity of both parents is transferred to the progeny. As a consequence, meiotic tetraploids are genetically more diverse and often display higher vigor than mitotic tetraploids (Parrot and Smith 1986). This is expected as red clover is an outbreeding species with a strong inbreeding depression (Boller et al. 2010), and plants that are highly heterozygous often perform better. One can therefore argue that genetic diversity will have a positive effect on seed yield, but this influence has been hardly investigated in the past. Indirect evidence was provided by Amdahl et al. (2015), who found that tetraploid cultivars created by crossing existing tetraploid cultivars tend to have higher seed yields than tetraploid cultivars derived from newly tetraploidized material (Amdahl et al. 2015).

Given the large variability for seed yield in red clover, a fair comparison of the performance of meiotic and mitotic tetraploids is ideally based on plant materials with similar genetic background. For example, meiotic tetraploids can be created by crossing diploid plants from a given cultivar with tetraploid plants from a different cultivar (Parrot and Smith 1986). Mitotic tetraploids with a similar background can be obtained by crossing tetraploidized plants from the first cultivar with tetraploid plants from the second cultivar. Flow cytometry is a suitable means to discriminate diploid and tetraploid progeny plants, while progeny plants that are the result of cross-pollination can be identified with molecular markers.

The aim of this study was to investigate seed yield in mitotic and meiotic tetraploids with similar genetic background. First, we generated suitable plant material. Second, we compared the seed yield of mitotic and meiotic tetraploid progeny plants for which we could confirm that they were the result of pair-crossing and explored the relation between genetic diversity and seed yield.

2 Materials and Methods

Two diploid cultivars Lemmon (LEM) and Merian (MER) known to produce unreduced gametes (Lamote 1998) were chosen. Tetraploid Lemmon and Merian populations (LEM^{4x} and MER^{4x}) were available (first generation after colchicine treatment). The tetraploid cultivar Elanus (EL) (Agroscope) was chosen as crossing partner because it is a well-performing tetraploid from a different origin. LEM, MER, LEM^{4x} and MER^{4x} plants were crossed with EL plants in 40 pair crosses (different genotypes in each cross) in insect proof gaze cages (85 × 85 × 110 cm) in a gaze tunnel greenhouse. During flowering, around 5 *Bombus terrestris* drones (Biobest, Belgium) were released weekly in each cage. Seeds were harvested on each individual plant at maturity.

From the successful interploidy crosses, 1815 large seeds from the LEM or MER parent and 576 seeds from the EL parent (4 crosses) were sown in Quickpot® trays (HerkuPlast, QP96T). From the tetraploid crosses, 576 seeds from the LEM^{4x} or MER^{4x} parent were sown analogously (Table 1). Seedlings were analyzed with flow cytometry (FC) to identify tetraploids. To confirm if tetraploid progeny plants were the result of cross-pollination, we used a multiplex SSR assay targeting 9 loci over all linkage groups, identified by Sato et al. (2005): RCS0883, RCS1501, RCS1679, RCS1647, RCS2728, RCS5376, RCS3236, RCS1868, RCS1897. DNA extraction,

SSR amplification and processing were carried out as described by Vleugels et al. (2014). PCR reactions were done in a total volume of 10 µl: 5 µl PCR Master Mix 2X, 0.36 µl primer mixture (0.02 µl per primer, 100 µmol/l), 2.64 µl H_2O and 2 µl DNA (10 ng/µl). PolyPatEx (Zwart et al. 2016) was used to identify tetraploid progeny from cross-pollinations using R v.3.2.2. (Found. Stat. Comp. 2015). The degree of self-pollination was calculated for each parent as the number of seedlings from self-pollinations over the total number of seedlings analyzed. When diploid parents from interploidy crosses produced diploid progeny, self-pollination was assumed..

Table 1. Summary of results for 7 tetraploid crosses. The number of seeds harvested on each parent, determination of diploid (2x) and tetraploid (4x) progeny with FC, progeny from cross-pollinations (CP) and the degree of self-pollination (SP) determined with SSR marker analysis.

Pair cross	Nr. of seeds		FC analysis			SSR analysis	
	LEM^{4x} MER^{4x}	EL^{4x}	Seedlings[a] analyzed	2x	4x	4x from CP^b	SP (%)
LEM^{4x}xEL1	84	569	157	0	154	10	91.7
LEM^{4x}xEL2	19	11	11	0	9	9	0.0
MER^{4x}xEL1	791	3385	61	0	59	56 (5)	4.9
MER^{4x}xEL2	684	398	60	0	57	44 (5)	21.7
MER^{4x}xEL3	4425	1569	53	0	37	19 (4)	34.0
MER^{4x}xEL4	195	4839	57	0	50	49 (5)	1.8
MER^{4x}xEL5	327	1562	62	0	57	42 (4)	24.2
Total	6525	12,333	461	0	423	229 (23)	42.1

[a]Triploid seedlings or seedlings with undeterminable ploidy are not listed.
[b]Progeny indicated in bold were tested for seed yield. The number of MER^{4x}xEL progeny used in the seed yield trial was adapted (brackets) to obtain similar numbers of LEM^{4x}xEL and MER^{4x}xEL progeny in the trial.

The seed yield of meiotic tetraploids (9 LEMxEL and 5 MERxEL) was compared to that of mitotic tetraploids (19 LEM^{4x}xEL and 23 MER^{4x}xEL) under field conditions. Plants were planted on 10/5/2016 in Merelbeke (Belgium; 50°58'37" N, 3°47'5"E, 19 m a.s.l.) in sandy loam soil in an alternated design. The trial was fertilized with 12 kg N/ha, 24 kg P_2O_5/ha and 160 kg K_2O/ha at planting. Weeds were controlled with 0.1 l/ha AZ500 at planting, 1.25 l/ha Corum on 16/6/16, and hoeing on 12/7/2016 and 28/7/16. During flowering all plants were loosely tied around a stick to prevent lodging. Seeds were harvested per plant at maturity: between 16/8/2016 and 2/9/2016. Seed yield data were analyzed with ANOVA, and Pearson correlations were calculated between seed yield and the average number of alleles per locus over 9 SSR loci using Statistica v.13 (Statsoft Inc., Tulsa, OK, USA).

3 Results and Discussion

Production of mitotic and meiotic tetraploids

Of the total of 40 pair-crosses performed, seeds were obtained from 16 interploidy crosses and 7 tetraploid crosses. For the remaining 17 pair-crosses either one of the parent plants perished before seed setting, or no seeds were formed.

From the tetraploid crosses, 461 seedlings from the LEM^{4x} or MER^{4x} parent were tested: 423 were tetraploid and 229 of these were the result of cross-pollination (Table 1). The lower number of effective tetraploids after FC analysis is due to unclear FC data or the occurrence of triploids.

From the interploidy crosses, 28 out of 1270 seedlings derived from diploid LEM or MER parent were tetraploid (2.2%), but only 12 of these were the result of cross-pollination (Table 2). Some progeny were triploid, had indeterminable ploidy level with FC or unclear SSR data. As expected, EL parents rendered mainly tetraploid progeny: 449 out of 474 seedlings (94.7%), but only 3 of them were the result of cross-pollination, indicating that the MER parent in cross MERxEL2 was able to produce tetraploid male gametes. This was not the case in other three pair-crosses for which seedlings from the EL parent were investigated. In total, 15 meiotic tetraploids were obtained from interploidy crosses.

The success rate for producing meiotic tetraploids was very low and independent of the ploidy of the parent: 0.7% (diploid) and 0.5% (tetraploid). However, collecting seed from the diploid parent has the advantage that numerous unwanted diploid or triploid seedlings could be discarded after FC, while collecting seed from the tetraploid parent results in numerous tetraploid seedlings that have to be analyzed with SSR markers which is much more expensive than FC. Future research should evaluate if there are more efficient strategies to generate meiotic tetraploids, such as top crosses.

The degree of self-pollination was considerable in our trial. In the 16 interploidy crosses, self-pollination was highly similar in diploid and tetraploid parents: 95.0% and 94.1% respectively (Table 2). Apparently, red clover plants almost exclusively self-pollinated, in the absence of a crossing partner with the same ploidy level. More strikingly, also in the tetraploid crosses where two tetraploid plants were present, self-pollination was relatively high: 42.1% on average, and exceeding 20% in 4 out of 7 mother plants (Table 1). This suggests that self-pollination may occur more than expected in tetraploids. Whether this is also the case in diploids cannot be concluded from these data as no diploid crosses were performed.

Seed yield in mitotic and meiotic tetraploids, in relation to genetic diversity

A total of 14 meiotic and 45 mitotic tetraploids were compared for seed yield per plant. Seeds were harvested on 53 individual plants (Fig. 1). Meiotic tetraploids yielded on average more seeds (15.9 g) than mitotic tetraploids with similar genetic background (10.4 g) (ANOVA p = 0.012). Our results confirm the observation of Amdahl et al. (2015) who found that tetraploid cultivars made by crossing existing tetraploids tend to have higher seed yields than newly tetraploidized cultivars. However, the mitotic tetraploids in our study were the first generation after colchicine treatment and it has been shown that first generation plants are often weaker than the subsequent generations after colchicine treatment (Taylor and Quesenberry 1996). This might be due to a

Table 2. Summary of results for 16 interploidy crosses. The number of seeds harvested on each parent, determination of diploid (2x) and tetraploid (4x) progeny with FC, progeny from cross-pollinations (CP) and the degree of self-pollination (SP) determined with SSR marker analysis.

Pair cross	Nr. of seeds		FC analysis			SSR analysis	
	LEM^{2x} MER^{2x}	EL^{4x}	Seedlings[a] analyzed	2x	4x	4x from CP^b	SP (%)
LEMxEL1	154	32	19	17	0	0	89.5
LEMxEL2	482	546	117	114	2	1	97.4
LEMxEL3	107		71 (LEM)	55	8	8	77.5
		985	236 (EL)	4	226	0	95.8
LEMxEL4	40	87	7	5	0	0	71.4
LEMxEL5	83	187	14	14	0	0	100.0
LEMxEL6	22	10	2	2	0	0	100.0
LEMxEL7	14	221	1	0	0	0	0.0
MERxEL1	1717	2500	451	445	6	1	98.7
MERxEL2	425		194 (MER)	187	3	0	96.4
		142	85 (EL)	1	77	3	87.1
MERxEL3	1493	1699	79	73	4	1	92.4
MERxEL4	1759	16	67	55	1	0	82.1
MERxEL5	173		131 (MER)	127	2	0	96.9
		97	70 (EL)	1	69	0	98.6
MERxEL6	25	279	2	1	0	0	50.0
MERxEL7	39		32 (MER)	31	1	1	96.9
		189	83 (EL)	5	77	0	92.8
MERxEL8	178	1183	61	60	1	0	98.4
MERxEL9	146	23	22	21	0	0	95.5
Total	6857		1270 (L/M)	1207	28	12	95.0
		8196	474 (EL)	11	449	3	94.1

[a]Triploid seedlings or seedlings with undeterminable ploidy are not listed.

[b]Progeny indicated in bold were tested for seed yield. The number of MER ^{4x}xEL progeny used in the seed yield trial was adapted (brackets) to obtain similar numbers of LEM ^{4x}xEL and MER^{4x}xEL progeny in the trial.

relatively high level of homozygosity and inbreeding depression. Possibly, mitotic tetraploids would have higher yields when more generations of multiplication had been performed.

Contrary to our expectations, meiotic tetraploids carried on average fewer alleles per locus (2.6) than mitotic tetraploids (2.9). In addition, seed yield and average allele number per locus were not correlated (p = 0.793), suggesting that seed yield and genetic diversity are not related. These results should be interpreted with care, however, because seed was harvested on only 12 meiotic tetraploids, and 9 SSR loci may be insufficient to provide an accurate estimate of genetic diversity. Further research should concentrate on the generation of a larger set of mitotic and meiotic tetraploids and the screening of a larger set of loci to estimate genetic diversity.

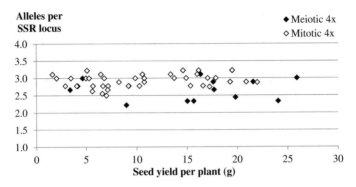

Fig. 1. Seed yield and average number of alleles per locus for 12 meiotic and 41 mitotic tetraploids with similar genetic background (LEM/MERxEL vs. LEM^{4x}/MER^{4x}xEL)

4 Conclusion

Self-pollination was common in interploidy crosses (95.0% to 94.1%). The combination of FC with SSR marker analysis to identify meiotic tetraploids among the progeny from the diploid parent was a cost-efficient strategy. The efficiency of interploidy crosses to generate tetraploids was low (0.5% to 0.7%). Meiotic tetraploids yielded significantly more seeds than mitotic tetraploids with similar genetic background (15.9 g vs. 10.4 g). Contrary to our expectations, we did not find any indication that meiotic tetraploids were genetically more diverse than mitotic tetraploids. Finally, no relation was found between genetic diversity and seed yield.

Acknowledgments. We thank Marianne Malengier for breeding work, Nancy Mergan, Michelle De Meyer and Hilde Carlier for the lab work, and the ILVO field technicians for maintaining the field trials.

References

Amdahl H, Aamlid TS, Ergon A, Rognli OA (2015) Seed yield of Norwegian and Swedish tetraploid red clover (*Trifolium pratense* L.) populations. Crop Sci 56(2):603–612

Boller B, Schubiger FX, Kölliker R (2010) Red clover. In: Boller B (ed) Handbook of plant breeding. Springer, Dordrecht, pp 439–455

Lamote V (1998) Opsporen van meïotische polyploïden en inductie van haploïden bij *Lolium* en *Trifolium* spp. PhD Thesis. Ghent University. p 111

Parrot WA, Smith RR (1986) Recurrent selection for 2n pollen formation in red clover. Crop Sci 26:1132–1135

Sato S, Isobe S, Asamizu E, Ohmido N, Kataoka R (2005) Comprehensive structural analysis of the genome of red clover (*Trifolium pratense* L.). DNA Res 12:301–364

Taylor NL, Quesenberry KH (1996) Red Clover Science. Kluwer Academic Publishers, Dordrecht, p 226

Vleugels T, Cnops G, Roldán-Ruiz I (2014) Improving seed yield in red clover through marker assisted parentage analysis. Euphytica 200(2):305–320

Zwart AB, Elliott C, Hopley T, Lovell D, Young A (2016) Polypatex: an R package for paternity exclusion in autopolyploids. Mol Ecol Resour 16(3):694–700

Domesticating Shrubby Native Legumes for Pastures and Natural Grasslands

J. P. Muir[1(✉)], F. S. Forrest[2], J. Brady[1], J. C. Dubeux Jr.[3], and W. D. Pitman[4]

[1] Texas A&M AgriLife Research, College Station, USA
j-muir@tamu.edu
[2] South Texas Natives, Texas A&M Kingsville, Kingsville, USA
[3] Institute for Food and Agricultural Science,
University of Florida, Gainesville, USA
[4] Louisiana State University Agricultural Center, Baton Rouge, USA

Abstract. As ruminant species diversify in natural and managed ecosystems in an effort to increase production, so does the need for greater plant diversity. Legume shrub propagation is rarely targeted for this purpose. As an example, the wide range of native perennial, shrub (0.5–3 m) legumes in Texas indicates untapped potential for multiple uses such as forage, ecosystems services, wildlife habitat, fuel, and possibly pulse crops. We propose that concerted programs to identify, collect, characterize, domesticate and commercialize these legumes could lead to multiple benefits in native and cultivated ecosystems characterized by herbaceous, shrubby and arboreal canopies, each contributing to greater herbage and animal productivity than mono-canopy systems. This approach could be applied in Europe and throughout the world.

Keywords: Fodder · Forage · Pasture · Perennials · Pulses

1 Introduction

Fostering greater diversity in grasslands results in enhanced productivity and stability. This has long been recognized in natural rangeland, especially in Africa where diverse vegetation has encouraged herbivore specialization, most notably specialization as grazers or browsers (Hofmann and Stewart 1972). Balancing domesticated browsers and grazers in rangelands has been historically used throughout the world to increase productivity and mitigate weed invasion (Warren et al. 1984). For example, Estell et al. (2014) defended the greater use of rangeland shrubs for livestock production.

Fodder diversity in cultivated pasture, by comparison, is only more recently being promulgated for enhanced productivity (Sanderson et al. 2007). This includes growing interest in silvopasture to grow arboreal and herbaceous fodder in the same paddock (Nair et al. 1984), especially legumes (Dubeux Jr. et al. 2014). Raising multiple domesticated herbivore species on pastures, beyond the occasional bovine/ovine/caprine combinations, is less common (Muir et al. 2015).

We propose that the addition of a legume shrub canopy in cultivated pasture or degraded natural grasslands will increase plant:herbivore stability and productivity.

G. Brazauskas et al. (Eds.): *Breeding Grasses and Protein Crops in the Era of Genomics*, pp. 46–50, 2018.
https://doi.org/10.1007/978-3-319-89578-9_8

Very few shrubby (0.5 to 3.0 m) fodder legumes have been domesticated for temperate and sub-tropical systems and greater effort could focus on this underutilized pasture and rangeland canopy. As an example, we briefly describe legume genera available in Texas and potential benefits of exploring this germplasm for domestication.

2 Materials and Methods

Taking native leguminous fodder species from natural ecosystems to commercialized seed is a long, complex process. There are successful examples of this for native grassland restoration germplasm domestication as undertaken by South Texas Natives (Smith et al. 2010) that provide some insights. We reviewed these approaches and outline the basic procedures for applying these to native leguminous shrub domestication, using the southern Great Plains of North America as a case study.

2.1 Plant Material Collection and Phenotyping

In most regions, pasture scientists are likely unaware of all the native shrub legume germplasm available in their region. This should not be surprising because their focus has historically been on introduced, high performance herbaceous species from around the world while local botanists tend to describe but not evaluate agronomic potential (Diggs Jr. et al. 1999). The first step, then, is to review the taxonomic literature for the target region or ecosystem. Table 1 below lists the genera in the southern Great Plains, mostly Texas, which fit into the general native legume shrub category, along with their published uses. Once promising shrub species in our target region have been identified, accession collections should then provide as wide a range of that species as possible. Genomic characterization, i.e. genetic variation within each species, should guide the narrowing or expansion of collection scopes. Examples of this process in herbaceous legumes include *Desmanthus illinoiensis* (Michx) MacM. (Kulakow et al. 1990) and *Acacia angustissima* (Mill.) Kuntze var. *hirta* (Nutt.) B.L. Rob. (Noah et al. 2012).

2.2 Agronomic Characterization

Shrub germplasm characterization could begin with seed germination trials to overcome dormancy (Dittus and Muir 2010). Field plot accession evaluation in multiple locations within the native range of the target species should follow, characterizing phenotypic growth, reproduction, diseases and pests. Previous South Texas Ecotype efforts with herbaceous legumes provide concrete examples of how this is accomplished (Smith et al. 2010). Microplot agronomic trials can then define herbage and seed yields, nutritive value, ligno-cellulose potential along with whatever else might be important (Noah et al. 2012). Ecosystems services, wildlife value and pulse potential could occur during this phase (Kulakow et al. 1990). Browser response to shrubs is the last, but not least, field trial focus.

Wider field evaluation under real-world conditions (pasture, rangeland or native ecosystem restoration) should be considered once promising shrubs have been identified. These might include edapho-climatic adaptation range under browsing pressure.

Table 1. Legume (Fabaceae) genera native to the southeastern USA with potential for browse domestication.

Genus	States	Known uses
Acacia	Texas, New Mexico	Browse, honey
Amorpha	Various	Browse, dye, ornamental
Baptisia	Texas	
Bauhinia	Texas	Horticulture
Brongniartia	Texas	
Caesalpinia	Texas, Mexico	Ornamental, tannin
Calliandra	Texas	Browse, fodder (seed)
Cercidium/Parkinsonia	Texas	Wood, ornamental, fuel, fodder (seed)
Cercis	Texas	Ornamental, fodder (seed), honey
Dalea	Various	Browse
Desmanthus	Texas, Mexico	Browse
Desmodium	Texas	Browse
Erythrina	Texas	Poison, wood
Eysenhardtia	Texas	Dye
Genistidium	Texas	
Gleditsia	Texas	Browse, ornamental, fodder (seed)
Krameria	Texas	Dye, browse, medicinal
Leucaena	Texas, Mexico	Browse, ornamental
Mimosa	Texas, Mexico	
Pithecellobium	Texas	Horticulture, furniture, browse, honey
Prosopis	Texas	Fuel, pulse, honey, fodder (pod), wood
Robinia	Texas	Browse, fodder (seeds)
Sesbania/Glottidium	Texas	
Sophora	Texas	Ornamental, browse, fodder (pod), medicinal (poison)
Wisteria	Texas, Louisiana, Arkansas	Ornamental

Turner 1959; Vines 1960; Correll and Johnston 1970; Diggs Jr. et al. 1999; Lehman et al. 2005

This is the most time consuming and costly evaluation stage but important in understanding herbivore/shrub interactions such as palatability and diet selection from the animal perspective as well as plant regrowth and persistence (Provenza et al. 2003; Sorensen et al. 2006).

2.3   Commercialization

Dissemination of those legume shrubs with agronomic potential is essential to getting germplasm into pastures, rangeland and restored ecosystems where they can contribute to diversity, stability and productivity. South Texas Natives has partnered with private landowners, agencies and commercial seed companies to increase and eventually sell native herbaceous legumes (Smith et al. 2010). This requires licensing production of germplasm to minimize chances of loss of seed company investment. This may raise some polemic because no genetic modification or breeding has occurred to this native germplasm; however, these materials do represent distinct subsets of plant materials, with empirical test results. Then foundation seed must be produced under genetic purity and weed-free conditions. A market must exist or be developed prior to further investment by seed companies. Subsidized demonstration projects of potential uses, in cooperation with peers, research stations, and agencies is usually necessary to stimulate market acceptance of new seed sources. This process may take years.

3   Conclusion

Diverse native leguminous shrubs exist in most regions and can serve as a basis for adding a fodder canopy to pastures and rangelands. This will require time and funds before this germplasm is widely available for widespread sale and use. On average, 5–7 years from wild seed collection to substantial consumer seed availability is required. Despite these limitations, the eventual increases in plant diversity, animal productivity, environmental benefits as well as financial returns make this an attractive proposition for cultivated pastures and degraded natural rangelands worldwide.

References

Correll DS, Johnston MC (1970) Manual of vascular plants of Texas. Texas Research Foundation, Renner

Dubeux JC Jr, Lira MA, dos Santos MVH, Muir JP, de Barros DAGM, da Silva MA, Teixeira VI (2014) Soil characteristics under legume and non-legume tree canopies in signal grass (*Brachiaria decumbens* Stapf.) pastures. Afr J Range Forage Sci 31:37–42

Diggs GM Jr, Lipscomb BL, O'Kennon RJ (1999) Shinners & Mahler's illustrated flora of North Central Texas. Botanical Research Institute of Texas, Fort Worth

Dittus D, Muir JP (2010) Scarification of native Texas legume seeds. Native Plants J 11:5–10

Estell RE, Havstad KM, Cibils AF, Anderson DM, Schrader TS (2014) The changing role of shrubs in rangeland-based livestock production systems: can shrubs increase our forage supply? Rangelands 36(2):25–31

Hofmann RR, Stewart DRM (1972) Grazer or browser: a classification based on the stomach structure and feeding habits of East African ruminants. Mammalia 36:226–240

Kulakow PA, Benson LL, Vail JG (1990) Prospects for domesticating Illinois bundleflower. In: Janick J, Simon JE (eds) Advances in new crops. Timber Press, Portland, pp 168–171

Lehman RL, O'Brien R, White T (2005) Plants of the Texas Coastal Bend. Texas A&M University Press, College Station

Muir JP, Pitman WD, Foster JL, Dubeux JC Jr (2015) Sustainable intensification of cultivated pastures using multiple herbivore species. Afr J Range Forage Sci 32:97–112

Nair PK, Fernandes ECM, Wambugu PN (1984) Multipurpose leguminous trees and shrubs for agroforestry. Agroforestry Syst 2:145–163

Noah RL, Muir JP, Brady JA, Wittie RD, Kattes DH, Pitman WD, Rea GL, Brakie MR (2012) Genotypic and phenotypic variability in three prairie acacia accessions. Crop Sci 52:951–959

Provenza FD, Villalba JJ, Dziba LE, Atwood SB, Banner RE (2003) Linking herbivore experience, varied diets, and plant biochemical diversity. Small Ruminant Res 49:257–274

50   J. P. Muir et al.

Sanderson MA, Goslee SC, Soder KJ, Skinner RH, Deak A (2007) Plant species diversity, ecosystem function and pasture management—a perspective. Can J Plant Sci 87:479–487

Smith FS, Lloyd-Reilley J, Ocumpaugh WR (2010) South Texas Natives: a collaborative regional effort to meet restoration needs in south Texas. Native Plants 11(3):252–268

Sorensen JS, Skopec MM, Dearing MD (2006) Application of pharmacological approaches to plant-mammal interactions. J Chem Ecol 32:1229–1246

Turner BL (1959) The legumes of Texas. University of Texas Press, Austin

Vines RA (1960) Trees, shrubs and woody vines of the southwest. University of Texas Press, Austin

Warren LE, Uekert DN, Shelton JM (1984) Comparative diets of Rambouillet, Barbado, and Karakul sheep and Spanish and Angora goats. J Range Manage 37:172–180

Effects of Temperature on Growth During *in vitro* Embryo Culture of Field Bean (*Vicia faba* var. *minor* L.)

P. Ogrodowicz[1(✉)], M. Surma[1], T. Adamski[1], Z. Kaczmarek[1],
W. K. Święcicki[1,2], P. Stopyra[3], A. Kuczyńska[1], K. Krystkowiak[1],
and K. Mikołajczak[1]

[1] Institute of Plant Genetics, Polish Academy of Sciences, Poznań, Poland
`pogr@igr.poznan.pl`
[2] Poznań Plant Breeders Ltd., Tulce, Poland
[3] IHAR-PIB Group, Plant Breeding Strzelce Ltd., Co., Strzelce, Poland

Abstract. Attainment of homozygous lines in grain legumes is multi-year procedure. It may be shortened by applying the single seed descent technique under greenhouse conditions. Additionally, this technique can be combined with *in vitro* culture of embryos dissected from immature seeds. The aim of this study was to establish *in vitro* conditions for the culture of field bean embryos as the first step in the development of homozygous lines. Four temperature regimes were applied and their influence on embryo development was evaluated. Under *ex vitro* conditions plant survival was observed. Our results indicate a good potential of embryo culture to shorten the time required for obtaining homozygous lines of field bean.

Keywords: Homozygous lines · SSD technique · Embryo development

1 Introduction

Grain legumes are important crops in Central and Northern Europe because they are a good source of fat, protein and carbohydrates. Field bean contains a high protein content (about 25–37%) in seeds which makes this species a valuable feed ingredient. Although field bean is a facultative cross-pollinating species, homozygous lines should be developed for breeding new cultivars and this process takes several years. Generally, homozygosity can be attained by selfing for several generations. In breeding, this procedure has usually been applied under field conditions which allows obtaining one generation per year. The single seed descent method (SSD; Goulden 1941) supplemented with *in vitro* culture of embryos has been proposed to shorten the breeding cycle (Surma et al. 2011, 2013). This approach allows obtaining about four generations per year in pea and two to three in lupins (Surma et al. 2013). The aim of this study was to test the influence of temperature during *in vitro* culture of field bean embryos as the first step optimise a procedure that shorten generation cycles during the development of field bean homozygous lines.

© Springer International Publishing AG, part of Springer Nature 2018
G. Brazauskas et al. (Eds.): *Breeding Grasses and Protein Crops in the Era of Genomics*, pp. 51–55, 2018.
https://doi.org/10.1007/978-3-319-89578-9_9

2 Materials and Methods

Four field bean (*Vicia faba* L. var. *minor*) cultivars: Albus, Granit, L5487 and Olga - all bred in Poland - were grown in the greenhouse. Pods from plants were detached 21–28 days after flowering, when seeds were still green but fully developed. Seeds were sterilized by successive dips into ethanol 70% (3 min) and Javel solution containing 1.5% active chlorine (5 min), and then rinsed three times with sterilized water. Dissected embryos were cultured on MS medium (Murashige and Skoog 1962) in tubes (∅25 × 150 mm).

Four temperature treatments were applied for embryo culture: K1 - 8 °C, K2 - the first week 16 °C, then 8 °C; K3 - the first week 20 °C, then 20/18 °C; K4 - 16 °C during whole culture period. In each temperature treatment the first week of culture was in the dark and in the next weeks photoperiod 8/16 h (day/night) was maintained. Experiment was performed in three replications. In each replication 10 tubes with two embryos were used for each cultivar and temperature treatment.

After 7, 14 and 21 days of culture, embryo development was observed by measuring shoot and root length. Plants of 3–5 cm with well-developed roots (about 30 days of culture) were transferred *ex vitro* to pots filled with mixture of peat substrate (Klasmann-Deilmann TS 1) and sand in the ratio 3:1 and they continued their growth in a greenhouse. 30–40 plants of each cultivar from each treatment were planted. The number of plants that survived were observed at 30 days after planting and expressed as percentage in relation to initial number of planted plants.

Statistics: the data for shoot and root length in 7, 14 and 21 days of culture were statistically processed using 3-way analysis of variance (ANOVA), in which genotypes, temperature treatments and culture duration (term of measurement), as well as corresponding interactions were taken into account as the sources of variation. All factors were considered to have a fixed effect. The percent of survival plants was not statistically processed because only one observation for each treatment was available.

3 Results and Discussion

ANOVA analysis showed a highly significant effect of temperature treatments on the development of plantlets in *in vitro* culture (Table 1). The differences between genotypes as well as the effect of genotype × temperature interaction were also significant. Significant differences between terms of observations indicate successive growth of plants in *in vitro* culture, but significant G x C interaction suggests uneven growth rates of the tested cultivars (Table 1).

Lengths of shoots and roots of field bean cultured *in vitro* in four different temperature regimes (K1, K2, K3, K4) measured after 7, 14 and 21 days are presented in Fig. 1.

Among the tested genotypes Olga and Granit were observed to have a similar growth rate, whereas L5487 was characterized by slower shoot growth, and Albus by slower root growth. It was noted for all the studied cultivars that growth rate of plantlets was a faster at higher temperatures (K3, K4) (Fig. 1, Table 2).

Table 1. Three-way analysis of variance for shoot and root length of field bean plants developed in *in vitro* culture under various temperature regimes.

Source of variation	d.f.[#]	Mean square	
		Shoot length	Root length
Genotype (G)	3	102.28**	60.24**
Temperature (K)	3	153.25**	125.50**
Culture duration (C)	2	211.13**	152.06**
Interaction:			
G × K	9	50.98**	46.89**
G × C	6	20.01*	80.53**
K × C	6	482.12**	322.66**
G × K × C	18	8.28	29.11**
Error	96	6.66	9.19

[#]d.f. - degrees of freedom; *$P < 0.05$; **$P < 0.01$

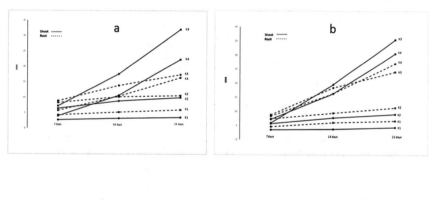

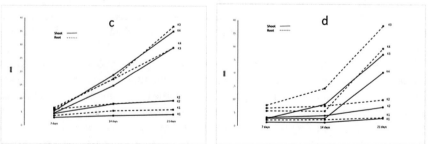

Fig. 1. Length of shoots and roots of field bean plantlets cultured *in vitro* in different temperature regimes (K1, K2, K3 and K4) measured after 7, 14 and 21 days; a – cv. Albus, b – cv. Olga, c – cv. Granit, d – cv. L5487; K1 - 8 °C; K2 - the first week 16 °C, then 8 °C; K3 - the first week 20 °C, then 20/18 °C; K4 - 16 °C during whole culture period.

Table 2. Shoot and root length of field bean plants developed in *in vitro* culture - mean values for cultivars, temperature conditions and days of culture duration.

Factor	Shoot length (mm)	Root length (mm)
Cultivar		
Albus	10.71	9.74
Olga	12.31	12.29
Granit	11.71	12.65
L5487	7.15	11.15
SE	0.608	0.714
Temperature regimes		
K1	3.11	4.69
K2	6.96	8.69
K3	17.02	17.63
K4	14.77	14.83
SE	0.608	0.714
Culture duration (days)		
7	4.36	6.30
14	9.41	10.58
21	17.63	17.50
SE	0.527	0,619

After potting, the differences between plants developed *in vitro* under various temperature treatments have decreased during cultivation. Moreover, *ex vitro* survival was higher for plants developed under low temperatures. Generally, *ex vitro* survival of plants was above 50% in most cases, ranging from 33.3% for Albus (K3) to 83.4% for L5487 (K1). Percentage of survived plants, averaged for all genotypes, amounted to 69.2% for K1, 64.6% for K2, 45.0% for K3 and 51.3% for K4.

A relatively high survival of field bean plants, compared, e.g., to lupins (Surma et al. 2013), may be due to the development of an abundant root system with numerous lateral and aerial roots, which was observed in field bean plantlets after 3–4 weeks of *in vitro* culture.

Low temperature treatments during *in vitro* culture of embryos was tested, based on the requirements of this species grown under field conditions. In early spring (end of March) seedlings of field bean are often exposed to low temperatures in the field. Plants that have experienced a period of low temperature, bloom abundantly and ripen early, which is desirable for using the SSD method. On the other hand, our studies reveal a genotype-temperature interaction effect, showing differences in the response to temperature across cultivars, and suggesting that cultivar-dependent optimization of culture conditions may be required.

4 Conclusions

In vitro culture of field bean embryos can be carried out in different temperature regimes. At 20 °C plantlets exhibit longer shoots and roots than those developed at 8 °C and 16 °C, but their *ex vitro* survival is lower. Results indicate that *in vitro* culture of embryos can complement the SSD technique conducted under greenhouse conditions.

Acknowledgements. The studies were supported by the Ministry of Agriculture and Rural Development, the National Multi-Year Project (2016–2020), Resolution no. 222/2015, 15 December 2015.

References

Goulden CH (1941) Problems in plant selection. In: Proceedings of the seventh genetics congress. Cambridge University Press, pp 132–133

Murashige T, Skoog F (1962) A revised medium for rapid growth and bioassays with tobacco tissue cultures. Physiol Plant 15:473–497

Surma M, Adamski T, Święcicki W, Barzyk P, Kaczmarek Z, Kuczyńska A, Krystkowiak K, Mikołajczak K, Ogrodowicz P (2013) Preliminary results of *in vitro* culture of pea and lupin embryos for the reduction of generation cycles in single seed descent technique. Acta Soc Bot Pol 82(3):231–236

Surma M, Adamski T, Kuczyńska A, Krystkowiak K, Trzeciak R, Mikołajczak K, Ogrodowicz P (2011) Zmodyfikowana technika pojedynczego ziarna w hodowli jęczmienia ozimego (Modified single seed descent technique in winter barley breeding). Biul IHAR, IHAR 260/261:59–66

Ecological and Agronomic Evaluation of *Dactylis glomerata* subsp. *lobata*

E. Norkevičienė[1(✉)], A. Balsevičius[2], V. Kemešytė[1], B. Butkutė[1], and G. Petrauskas[1]

[1] Institute of Agriculture, Lithuanian Research Centre for Agriculture
and Forestry, Kėdainiai, Lithuania
egle.norkeviciene@lammc.lt
[2] Botanical Garden of Vytautas Magnus University, Kaunas, Lithuania

Abstract. Widening of ecological adaptation of forage grasses (including tolerance of climatic stresses, enhancement of genetic resistance to pests and diseases, expansion of geographical adaptability) warrant exploration and utilization of wild relatives of a wide range of species. This study set out to investigate the ecological characteristics of *Dactylis glomerata* subsp. *lobata* and to compare its agronomic traits and yield with those of the most common and economically valuable species of the genus, *D. glomerata* subsp. *glomerata*. The current study revealed that *D. glomerata* subsp. *lobata* is less productive than *D. glomerata* subsp. *glomerata*. Though, *D. glomerata* subsp. *lobata* agrobiological characteristics suggest that it might be a valuable genetic resource for forage production.

Keywords: *Dactylis glomerata* subsp. *lobata* · Genetic resources
Forage yield · Agronomic traits

1 Introduction

The genetic base of sown pastures in the European Union is still quite narrow. FAO highlights the need to widen the choice of the currently available high-value grass and legume cultivars by exploring, evaluating and selecting from a wide range of species of several genera (Batello et al. 2008). In recent years, there has been an increasing interest in wild relatives of cultivated plants because they provide a broad pool of potentially useful genes (Muñoz et al. 2017). Forage breeding of *Dactylis* genus species has been very intense in recent decades as it is the fourth most economically important forage grass genus grown worldwide (Stewart and Ellison 2010). Along with *D. glomerata,* there is elevated interest in *D. glomerata* subsp. *lobata* (Drejer) H. Lindb. Unlike tetraploid ($2n = 4x = 28$) *D. glomerata* subsp. *glomerata*, subspecies *D. glomerata* subsp. *lobata* is diploid ($2n = 2x = 14$). Gene flow between the two subspecies has been reported to occur in nature (Lumaret and Barrientos 1990). Some authors consider it as an attractive forage-type grass (Dabkevičiene et al. 2007; Mika et al. 2003). Dabkevičiene et al. (2007) have reported that *D. glomerata* subsp. *lobata* tend to provide better forage quality than *D. glomerata*. However, it has still not been comprehensively characterized and generally domesticated. The study seeks to evaluate

© Springer International Publishing AG, part of Springer Nature 2018
G. Brazauskas et al. (Eds.): *Breeding Grasses and Protein Crops in the Era of Genomics*, pp. 56–61, 2018.
https://doi.org/10.1007/978-3-319-89578-9_10

the ecological characteristics of *D. glomerata* subsp. *lobata* and to compare its agronomic traits with those of *D. glomerata* subsp. *glomerata* under Lithuanian conditions.

2 Materials and Methods

The vegetation database PHYTOSUD, developed by the authors of the paper in 1992, was used for the assessment of geographical and ecological characteristics of *D. glomerata* subsp. *lobata*. Syntaxa names follow Mucina et al. (2016).

The annual mean precipitation at the experimental site is 550–600 mm and mean annual temperature is 6.0–6.5 °C. The mean January temperature during 2012–2014 was −5.3 °C and the mean July temperature was 19.1 °C. Normally, average temperature in January, the coldest month in Lithuania, is −2.9 °C, and in the warmest month of July is 19.7 °C. However, during the 2012–2014 experimental period the weather conditions markedly differed between years. The first season of 2012–2013 was characterized by a cold and long-lasting winter, very late, dry and warm spring and changeable summer with heat waves. The autumn–winter period of 2013–2014 was unusually warm as the prevalent temperatures remained positive, i.e., the average monthly temperatures in January and February were more than 5 °C higher than the long-term mean temperature.

Field trials were carried out during 2012–2014 at the Institute of Agriculture, Lithuanian Research Centre for Agriculture and Forestry. The soil of the experimental site is *Endocalcari-Epihypogleyic Cambisol*. Seeds of a *D. glomerata* subsp. *lobata* accession originated from Kaliningrad region (54° 30′ 81″ N; 20° 00′ 42″ E) and late maturing *D. glomerata* cultivar 'Aukštuolė' were sown in a field trial in plots of 14.68 m^2. Sowing rate was determined taking into account seed germination: *D. glomerata* subsp. *lobata* – 12.6 kg ha^{-1}, *D. glomerata* cultivar 'Aukštuolė' – 11.21 kg ha^{-1}. The trial was laid out in a randomized complete block design with 4 replications for each subspecies. Pre-planting fertilization comprised of nitrogenous, phosphorus and potassium fertilisers ($N_{15}P_{45} K_{90}$) were applied before the sowing. Nitrogenous fertilisers (N_{190}) were applied each year of herbage use in several applications: in spring N_{70}, and N_{60} after the first and second cut.

The following agrobiological traits were examined: winter survival (%), early spring growth (1–9 points), plant height (cm, at inflorescence emergence), inflorescence length (cm), heading date (day of year recorded when 50% of plants are flowering), dry matter yield (t ha^{-1}). In order to determine dry matter yield the grasses were harvested three times per season: at the beginning of heading; after re-growth of the first and the second aftermath, considering at each subspecies specific date of maturity stage. Dry matter yield was determined by drying 500 g herbage samples at 105 °C to constant weight. In order to determine dry matter structure 0.5 kg herbage subsample of the first cut of the plots in each replication were collected. The herbage samples were dried at 60 °C to constant moisture. Two samples (50 g each) from replication were analysed assessing stem, inflorescence and leaf weight (g).

ANOVA were performed to test the effects of subspecies and year on evaluated traits. After a significant effect was detected, Fisher's LSD test was used to test the significance of differences between the treatment means. Differences were considered to be significant at the 95% level. Mean ± standard error (SE), standard deviation and coefficient of variation were used to describe the variability of measurements. Statistical analysis was conducted using the SAS Enterprise Guide 7.1 (SAS Institute Inc.).

3 Results and Discussion

3.1 Geographical and Ecological Characteristics of *Dactylis glomerata* subsp. *lobata*

The current International Code of Botanical Nomenclature indicates that this taxon has two accepted epithets (according to the rank in which it is used). When the taxon is determined as a species, the correct name is *D. polygama* Horv., whereas in the case of the subspecies it is *D. glomerata* L. subsp. *lobata* (Drejer) H. Lindb (synonym *Dactylis glomerata* subsp. *aschersoniana* (Graebn.) Thell.) (Miziantų 2008). *D. glomerata* subsp. *lobata* is a perennial herbaceous plant typical of shade-adapted forest species (Lambers et al. 2008). *D. glomerata* subsp. *lobata* is distributed mainly in Europe: from France in the West to Georgia in the East; from Italy and Turkey in the South to Sweden, Northwest Europe and Russia in the North.

In Lithuania, *D. glomerata* subsp. *lobata* is a rare and protected species, included in the Red Book of Lithuania (Čiuplys 2007). Its distribution and ecology are insufficiently investigated. The species mainly occurs in the southern part of the country, where it reaches the north-eastern border of its geographical distribution area. Nevertheless, several sites in North and Northeast Lithuania are also known (Čiuplys 2007). These sites are likely to be beyond the range of the main species distribution.

In the southwestern part of Lithuania, *D. glomerata* subsp. *lobata* mainly occurs in the zonal mesic nemoral forest communities of *Carpinion betuli* alliance (*Carpino-Fagetea* class) or in thermophilous oak forests (*Quercetea pubescentis* class). The populations growing on the edge or beyond the distribution range are found in zonal subboreal *Querco roboris-Tilion cordatae* alliance communities, which geographically substitute temperate *Carpinion betuli,* as well as they occur in ecologically different communities of *Alnion incanae* alliance (*Alno glutinosae-Populetea albae* class). This class includes riparian gallery forests of the Eurosiberian and Mediterranean, and the alliance includes alder-ash and oak riparian floodplain forests on nutrient-rich alluvial soils in the nemoral and subboreal zone of Europe (Mucina et al. 2016).

This information is essential to better understand the ecological peculiarities of *D. glomerata* subsp. *lobata* in order to conserve its genetic resources *in situ, ex situ* in Lithuania as well as to adjust growing conditions in the field. However, more research on this topic needs to be undertaken before the ecology of the wild ecotypes is more clearly understood in order to implement its domestication for forage production.

3.2 Agronomic Characteristics of Dactylis glomerata subsp. lobata

Results of this study revealed moderate differences for most of the tested traits between *D. glomerata* subsp. *lobata* and *D. glomerata* 'Aukštuolė' in both experimental years 2013 and 2014 (Table 1). *D. glomerata* subsp. *lobata* showed slightly lower winter survival and early spring growth compared with *D. glomerata* 'Aukštuolė'. However, according to other characteristics, the plants of the accession of *D. glomerata* subsp. *lobata* exhibited significantly ($P < 0.05$) shorter height, higher leaf number per plant and later maturity (2 to 7 days later flowering) than *D. glomerata* (Table 1, Fig. 1). These traits are desired for most appropriate pasture plant species. The highest values of variance for both species were identified for annual dry matter yield which is a complex trait, influenced by multiple factors.

Table 1. The variation of agronomic traits between *Dactylis glomerata* subsp. *lobata* and *D. glomerata* 'Aukštuolė' in the 1st (2013) and 2nd (2014) harvest years.

Trait	Mean[1]		Std. Dev.		Min		Max		Coef.Var.	
	Harvest year									
	1st	2nd	1st	2nd	1st	2nd	1st	2nd	1st	2nd
D. glomerata 'Aukštuolė'										
WS, %	98.6^a	95.8^a	3.5	7.5	88	77	100	100	3.5	7.8
ESG, point	9^a	7.9^b	0	0.9	9	7	9	9	0	11.7
PH, cm	126^a	89^b	6.9	4.2	115	80	135	95	5.4	4.8
IL, cm	13.3^{ab}	15.2^a	2.2	2.4	9.5	11	16	18	16.5	15.9
HD, DOY	139^a	136^a	0.4	5.8	139	122	140	139	0.3	4.2
DMY, t ha^{-1}	7.5^a	5.7^b	1.3	1.6	4.3	4	9.4	8	18.1	27.2
D. glomerata subsp. *lobata*										
WS, %	94.4^a	97.2^a	7.5	4.5	77	88	100	100	7.9	4.6
ESG, point	8.1^{bc}	6.1^d	0.9	0.5	7	5	9	7	11.8	9.2
PH, cm	82^c	77^d	1.5	2.8	79	72	85	82	1.8	3.6
IL, cm	11.8^b	14^{ab}	2.2	2.4	8	11	16	18	22.8	13.4
HD, DOY	141^b	143^c	1.1	1.1	140	143	143	146	0.8	0.7
DMY, t ha^{-1}	6.4^{bc}	5.2^{bd}	1.5	0.7	4.3	4.2	9.3	6.0	24	14.4

[1]Different letters indicate significant differences between subspecies and between years at the 0.05 probability level.
WS – winter survival, *ESG* – early spring growth, *PH* – plant height, *IL* – inflorescence length, *HD* – heading date, *DMY* – annual dry matter yield.

To our knowledge, there is no published data on the productivity of *D. glomerata* subsp. *lobata* in the northern regions of Europe. Our study suggests that total dry matter yield of *D. glomerata* subsp. *lobata* was significantly ($P < 0.05$) lower than that of *D. glomerata* in both experimental years (Fig. 2). The greatest difference was recorded in the first harvest taken at the beginning of heading – the yield of *D. glomerata* subsp. *lobata* was 25% lower compared to *D. glomerata* 'Aukštuolė'.

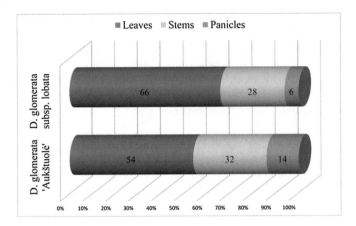

Fig. 1. The distribution of *Dactylis glomerata* subsp. *lobata* and *D. glomerata* 'Aukštuolė' biomass within plant structural components. Numbers in columns denote dry biomass yield per plant in %.

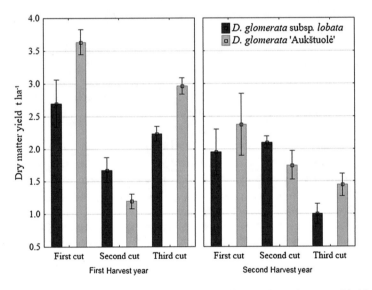

Fig. 2. Dry matter yield of *Dactylis glomerata* subsp. *lobata* and *D. glomerata* 'Aukštuolė' in the 1ˢᵗ (2013) and 2ⁿᵈ (2014) harvest years. Error bars indicate standard error.

4   Conclusion

In Lithuania, *Dactylis glomerata* subsp. *lobata* grows in the communities of *Carpinion betuli*, *Querco roboris-Tilion cordatae* and *Alnion incanae* alliances. Because of their adaptation to ecologically diverse environments, they might be a valuable genetic resource for forage breeding programs. However, the different ploidy level makes their use in *D. glomerata* subsp. *glomerata* breeding programs not straightforward. *D. glomerata* subsp. *lobata* exhibited lower biomass productivity, higher number of leaves, a better re-growth after the first harvest and later maturity than *D. glomerata* 'Aukštuolė'. Further research is warranted to ascertain the full potential and limitations of *D. glomerata* subsp. *lobata* for being used for forage as well for special purposes lawns production in Lithuania.

Acknowledgments. This study was funded through the long-term research program "Genetics and purposeful change of genotypes of agricultural and forest plants" implemented by Lithuanian Research Centre for Agriculture and Forestry.

References

Batello C, Mannetje L, Martinez A, Suttie J (2008) Plant genetic resources of forage crops, pasture and rangelands. Thematic background study. FAO report:5–7

Čiuplys R (2007) Miškinė šunažolė (*Dactylis polygama* Horv.). In: Rašomavičius V (ed) Lietuvos Raudonoji Knyga. Vilnius, p. 614

Dabkevičienė G, Paplauskienė V, Tarakanovas P, Lemežienė N, Liatukienė A (2007) Wild populations of *Dactylis polygama* H. for the formation of genetic collection and breeding. Biologija 53(3):12–15

Lambers H, Chapin SF III, Pons TL (2008) Plant physiological ecology, 2nd edn. Springer, New York

Lumaret R, Barrientos E (1990) Phylogenetic relationships and gene flow between sympatric diploid and tetraploid plants of *Dactylis glomerata* (*Gramineae*). Plant Syst Evol 169:81–96

Mika V, Kuban V, Kleidus B, Odstrčilova V, Hartmann J (2003) Phenolic spectra in genetic resources of grasses and legumes. Czech J Genet Plant Breed 39(spec. iss.):218–220

Mizianty M (2008) *Dactylis glomerata* subsp. *lobata* (*Poaceae*) w Polsce (uwagi nomenklatoryczne). Fragm Flor Geobot Pol 15(2):163–170

Mucina L, Bültmann H, Dierßen K, Theurillat JP, Raus T, Čarni A, Chytrý M (2016) Vegetation of Europe: hierarchical floristic classification system of vascular plant, bryophyte, lichen, and algal communities. Appl Veg Sci 19(S1):3–264

Muñoz N, Liu A, Kan L, Li MW, Lam HM (2017) Potential uses of wild germplasms of grain legumes for crop improvement. Int J Mol Sci 18(2):328

Stewart AV, Ellison N (2010) The Genus *Dactylis*. In: Wealth of wild species: role in plant genome elucidation and improvement, vol. 2. Springer, New York

Yield and Genetic Composition of Latvian ×*Festulolium* Cultivars and Breeding Material

P. Bērziņš[1], D. Ruņģis[2(✉)], S. Rancāne[1], A. Gailīte[2], V. Beļeviča[2],
V. Stesele[1], I. Vēzis[1], and A. Jansons[1]

[1] Institute of Agriculture, Latvia University of Agriculture, Skriveri 5101, Latvia
[2] Genetic Resource Centre, Latvian State Forest Research Institute "Silava",
Salaspils 2169, Latvia
dainis.rungis@silava.lv

Abstract. Interspecific hybridisation of forage grasses has been undertaken at the Institute of Agriculture since the end of the 1970's. *Lolium perenne (Lp)* and *Festuca pratensis (Fp)* aneuploid hybrids and a tetraploid *Fp* cultivar 'Patra' have been utilised for crossing. Two cultivars – 'Saikava' and 'Vizule', which contain *Lp*, *L.multiflorum (Lm)* and *Fp* germplasm have been registered. These *xFestulolium* cultivars were assessed in field conditions, and the genetic composition was determined by morphological parameters as well as by utilisation of species-specific molecular markers. In total, a higher proportion of *Lolium* (0/L) genotypes were identified in the analysed loci (62%), compared to hybrid genotypes (F/L – 34%). *Festuca* genotypes (F/0) were found in 3% of all loci. A higher proportion of *Lolium* alleles were found within all the *xFestulolium* cultivars: Ape – 74.9%, Lorry – 81.8%, Vizule – 72.9%, Saikava – 75.8%, Lofa – 74.3%, Punia – 58.9%.

Keywords: *Festuca* · *Lolium* · *xFestulolium*
Species-specific indel DNA markers

1 Introduction

Perennial ryegrass (*Lolium perenne* L.) is high yielding species with a very good digestibility. It is the dominant forage grass species in Europe due to its high regrowth capacity, rapid establishment, tolerance to frequent cutting and high nutritive value for ruminant livestock (Wilkins and Humphreys 2003). In Latvian conditions *Lp* often is not sufficiently winter-hardy, therefore interspecific hybridisation is performed by mutual pollination of *Lp* and *Fp*. Important requirements for *xFestulolium* cultivars is combining such characters of ryegrass as productivity, growth potential and forage quality, with those of fescues – stress tolerance such as cold and winter persistence during dormancy and drought during the growth period (Casler 2002).

Development of *xFestulolium* cultivars involves obtaining hybrids and renewal of fertility. In the Latvian forage grass breeding program, hybrids are obtained by the pollination by the desired pollen donor of spontaneously formed triploid or higher ploidy individuals (Berzinš et al. 2015). This method has allowed us to obtain hybrids between the major *Lolium* and *Festuca* species (*Lp*, *L.multiflorum (Lm)*, *Fp*, *F.arundinacea*,

© Springer International Publishing AG, part of Springer Nature 2018
G. Brazauskas et al. (Eds.): *Breeding Grasses and Protein Crops in the Era of Genomics*, pp. 62–66, 2018.
https://doi.org/10.1007/978-3-319-89578-9_11

F.gigantea, F.rubra). Fertility is restored by repeated crossings. The initial material utilised in hybrid formation contained cultivars with *Lm* germplasm, therefore our *xFestulolium* cultivars contain some contribution from *Lm*.

In order to select germplasm with the desired *Festuca* and *Lolium* proportions, it is important to evaluate hybrids by morphological and genetic analyses. The aim of the study was to evaluate *xFestulolium* cultivars in field conditions, and to assess their genetic composition using species-specific DNA markers.

2 Materials and Methods

Field Trials. Since 2007 to 2016 *xFestulolium* cultivars were evaluated in the exper-imental field of the LLU Institute of Agriculture in Skriveri (56°37 N and 25°07 E). Dry matter yield (DMY) and description of morphological traits such as culm formation in aftermath was carried out in trials established repeatedly from 2006 to 2015 in a sod-podzolic loam soil. Fertiliser with PK 60:90 kg ha^{-1} and N 120–180 kg ha^{-1} was applied annually.

Plant Material and Genotyping. Seeds were germinated on moistened filter paper in a growth cabinet (22 °C/20 °C, 16 h light) for approximately two weeks. DNA was extracted from leaf tissue using a CTAB-based method (Porebski et al. 1997). Fifteen indel-type EST-based markers that were previously reported to distinguish *Lp* and *Fp* were utilised (Os01g01080, Os01g34480, Os01g43070, Os03g50480, Os03g64210, Os04g06790, Os04g30420, Os04g54410, Os05g19630, Os05g24550, Os06g11040, Os06g41790, Os07g25430, Os10g25360, Os11g09280) (Tamura et al. 2009). Initially, markers were tested on 2 individuals of cultivars representative of *Fp* (Priekuļu 519), *Lp* (Spīdola), *Lm* (Taro (Denmark) and Fredrik (Sweden)). PCRs were performed in a total volume of 20 µl containing 2 µL DNA, 4 µl 5x HOT FIREPol® Blend Master Mix with 10 mM MgCl$_2$ (Soltis Biodyn, Tartu, Estonia), 1 µl of each primer (4 mM). The PCR program consisted of an initial denaturation step at 95 °C for 15 min; 40 cycles of denaturation at 95 °C for 20 s, annealing at 54 °C for 30 s, elongation at 72 °C for 45 s; a final elongation step at 72 °C for 5 min. PCR products where purified using a SephadexG50 spin plate, diluted 5x with water, and PCR fragments analyzed using a DNA 5K Labchip and HT DNA 5K reagent kit on a LabChip GX Touch HT instrument. The genotypes were analysed by dispersion analysis, and PCoA was done using GenAlEx (Peakall and Smouse 2012) after genotypes at each locus were coded as *Festuca, Lolium* or hybrid.

3 Results and Discussion

Due to poor WH, large areas of *Lp* were not grown in Latvia until the middle of the last century, when winter-hardy cultivars were developed – the diploid 'Priekuļu 59' (now 'Gunta') and later the tetraploid 'Spīdola'. In an effort to improve *Lp* properties, the yield and quality components of *Lolium* were combined with the WH of *Festuca*. Hybrids of these species tend to have a large variety of morphological forms and fertility levels. Currently, the most promising material with good WH was developed using crosses between the tetraploid *Fp* variety 'Patra' and *Lp* germplasm. Data obtained in Skriveri during 4 year period show that the relative DMY of *Fp* and *FL* were: 85–110% and 125–150% in the 1st ley year; 115–150% and 110–125% in the 4th ley year; 95–115% and 110–120% on average in 4 years, respectively.

In Latvia, both *Lolium* and *Festuca* have sufficient WH in the 1st winter after sowing. After that, depending on winter conditions, survival can be sufficiently high (70–90%), or it can be dramatically reduced. However, hybrids have good renewal, depending on the cultivar, and they can survive for 4–5 and more years in fields.

Culm formation data indicate the yield benefits contributed by *Lm*, unfortunately in combination with decreased WH (Table 1). The incorporation of *Festuca* into this material may increase WH without decreasing yield.

Table 1. The average DMY (t ha^{-1}) and formation of culms (CF) in aftermath

Variety	DMY 2007–08	CF 3.08.08	DMY 2010–12	DMY 2013–16	CF 9.08.16	DMY 2016	CF 12.07.16
Saikava (Latvia)	7.89	5.2	7.19	6.72	6.7	8.03	4.9
Vizule (Latvia)			7.33	6.81	6.8	8.53	4.8
Lofa (Czech Rep.)	7.92	8.1				9.29	7.8
Punia (Lithuania)	8.46	6.3	6.81	4.81	8.0		
LSD 0.05	0.72	2.14	0.60	1.38	0.59	0.78	0.85

CF - Culms in aftermath: 1 - none; 9 - intense formation

Based on the initial analysis, the marker Os01g34480 did not distinguish *Festuca* and *Lolium* varieties, and the markers Os07g25430 and Os11g09280 did not amplify clearly interpretable fragments. These markers were not utilised for further analysis. The remaining twelve markers were utilised to analyse twelve individuals from each of the six *xFestulolium* cultivars.

In total, a higher proportion of *Lolium* (0/L) genotypes were identified in the analysed loci (62%), compared to hybrid genotypes (F/L – 34%). *Festuca* genotypes (F/0) were found in 3% of all loci, while failed PCRs accounted for 1% of analysed loci. A higher proportion of *Lolium* alleles were found within all the cultivars: Ape – 74.9%, Lorry – 81.8%, Vizule – 72.9%, Saikava – 75.8%, Lofa – 74.3%, Punia – 58.9% (Table 2), as reported previously (Canter et al. 1999).

The *L* × *boucheanum* 'Saikava' was developed from 'Ape'. Both contain a genetic contribution from the hybrid variety 'Vizule', which has been maintained as a separate form from 'Ape' and 'Saikava', however these cultivars share a similar pedigree and have been genetically influenced by each other. The remaining three ('Lofa', 'Lorry', 'Punia') have distinct pedigrees and therefore they were expected to be more genetically distinct from 'Saikava', 'Vizule' and 'Ape'.

Table 2. Estimation of genetic composition using 12 markers in six cultivars

Cultivar	Similar[b]	Group[a]	0/L%	FL%	F0%	L%	F0 + FL%	Group[a]	0/0%
1. Ape	61.36 ± 1.73	5	61.8	34.0	0.7	74.9	34.7 ± 3.97	2	3.5
2. Saikava	69.44 ± 1.64	4	68.1	29.9	2.1	75.8	31.9 ± 3.89	3	0.0
3. Vizule	70.83 ± 1.62	3	62.5	36.1	1.4	72.9	37.5 ± 4.03	2	0.0
4. Lofa	68.81 ± 1.65	4	64.6	31.3	2.8	74.3	34.5 ± 3.95	2	1.4
5. Lorry	80.93 ± 1.40	1	77.1	21.5	1.4	81.8	22.9 ± 3.50	4	0.0
6. Punia	75.13 ± 1.54	2	37.5	51.4	11.1	58.9	62.5 ± 4.03	1	0.0
Average			61.9	34.0	3.2	74.1			0.8

FL - hybrid genotype, 0/L - *Lolium* genotype, F0 - *Festuca* genotype, 0/0 - failed PCR
[a]No differences between cultivars from the same group; [b]Similar between 12 individuals

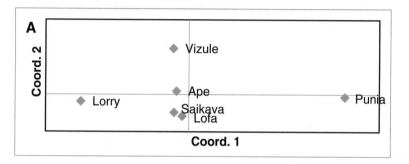

Variation explained by axis 1 – 81.72%, axis 2 – 10.83%

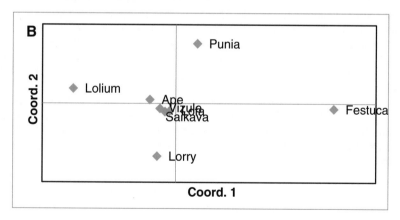

Variation explained by axis 1 – 94.01%, axis 2 – 4.41%

Fig. 1. PCoA of *xFestulolium, Festuca* and *Lolium* cultivars.

The genetic diversity within cultivars was investigated by analysing the number of loci that had identical genotypes within each cultivar (Table 2). 'Ape' was the most genetically polymorphic, while 'Lorry' and 'Punia' were genetically the most uniform.

Principal coordinates analysis (PCoA) of cultivars of the hybrid cultivars (Fig. 1a) and of the hybrid cultivars and *Lolium* and *Festuca* (Fig. 1b) illustrate that the hybrid cultivars are genetically more similar to *Lolium*. 'Punia' and 'Lorry' were differentiated from the other *xFestulolium* cultivars.

4   Conclusion

The yield of *xFestulolium* cultivars are significantly higher than those of *Lolium* – in the 1st year DMY is 125–150% compared to *Lolium*, while in the 4th year it is 110–120%. A higher proportion of *Lolium* alleles were found within all the analysed *xFestulolium* cultivars.

References

Bērziņš P, Rancāne S, Stesele V, Jansons A (2015) Current activities and challenges in breeding work with interspecific hybrids of perennial grasses. In: VI baltic genetic congress, Tartu University, Estonia, 30 September–3 October 2015, p 38

Canter PH, Pašakinskienė I, Jones RN, Humphreys MW (1999) Chromosome substitutions and recombination in the amphiploid *Lolium perenne* × *Festuca pratensis* cv Prior (2n=4x=28). Theor Appl Genet 98:809–814

Casler M (2002) Natural selection for survival improves freezing tolerance, forage yield and persistance of Festulolium. Crop Sci 42:1421

Peakall R, Smouse PE (2012) GenAlEx 6.5: genetic analysis in Excel. Population genetic software for teaching and research-an update. Bioinformatics 28:2537–2539

Porebski S, Bailey LG, Baum BR (1997) Modification of a CTAB DNA extraction protocol for plants containing high polysaccharide and polyphenol components. Plant Mol Biol Report 15:8–15

Tamura KI, Yonemaru JI, Hisano H, Kanamori H, King J, King IP, Tase K, Sanada Y, Komatsu T, Yamada T (2009) Development of intron-flanking EST markers for the *Lolium/Festuca* complex using rice genomic information. Theor Appl Genet 118:1549–1560

Wilkins PW, Humphreys MO (2003) Progress in breeding perennial forage grasses for temperate agriculture. J Agric Sci 140:129–150

Genetic Variability of the Most Important Traits in Alfalfa Cultivars Under Lithuanian Conditions

A. Liatukienė[(⊠)]

Institute of Agriculture, Lithuanian Research Centre
for Agriculture and Forestry, Kėdainiai, Lithuania
aurelija.liatukiene@lammc.lt

Abstract. Alfalfa (*Medicago sativa* L.) is an important perennial legume for animal feeding. A basic requirement for the successful breeding of this species is variability of the initial material. The collection under study consisted of 37 cultivars of different origin. The current study aimed to determine the genetic variability of the most important traits, including disease resistance, seed yield, fresh and dry matter yield, plant height during spring regrowth and at flowering stage and stem density. The statistical analysis of the data (ANOVA) revealed significant variability among the cultivars for all traits investigated. The highest variability was found for the seed yield (46.5%), while the lowest one (5.9%) was estimated for resistance to spring black stem and leaf spot. A cluster analysis was used to graphically illustrate the genetic diversity of alfalfa cultivars. The high level of genetic variability among the alfalfa cultivars in this collection highlights their potential for being used in the future breeding schemes.

Keywords: Alfalfa · Genetic variability · Breeding

1 Introduction

Alfalfa is one of the most widely cultivated perennial forage legumes. It is grown for hay, silage and grazing as a valuable crop characterised by a high nutritional quality, abundant biomass production, and broad adaptability to a wide range of cultivation conditions as well as by favourable environmental impacts. It exerts positive effects on soil fertility and structure and prevents soil erosion (Herrmann et al. 2010; Bouton 2012; Sabanci et al. 2013). However, deficiency of high complex disease resistance is one of the main constrains for successful cultivation of alfalfa (Lamb et al. 2006). The recent investigation of alfalfa for resistance to diseases in Lithuania suggests that a broad range of disease causal agents can heavily damage all parts of plants. Liatukienė and Žilvinas (2010) have reported that under wet Lithuania's climate conditions, fungal diseases are the key limiting factor for a stable and high seed and herbage yield and its quality. They are one of the reasons why alfalfa cultivation area in Lithuania and the whole Baltic Sea Region is still negligible (Anonymous 2012). However, under dry and hot climate conditions alfalfa yields much better (Rashidi et al. 2009). Several studies have documented that alfalfa material of different origin considerably differed in

© Springer International Publishing AG, part of Springer Nature 2018

G. Brazauskas et al. (Eds.): *Breeding Grasses and Protein Crops in the Era of Genomics*, pp. 67–72, 2018.
https://doi.org/10.1007/978-3-319-89578-9_12

resistance to Sclerotinia crown and stem rot (*Sclerotinia trifoliorum* Erikss.) (Kanbe et al. 2002), downy mildew (*Peronospora trifoliorum* de Bary.) (Jie et al. 2000; Yaege and Stuteville 2000), and spring black stem and leaf spot (*Phoma medicaginis* var. *medicaginis*) (Ellwood et al. 2006; Castell-Miller et al. 2007). The availability of genetically diverse alfalfa material enables development of cultivars with improved disease resistance (Nagl et al. 2011). The study of Lamb et al. (2006) evidenced that cultivars yielding improvement during 50 years of breeding were very environment dependent. The main advantage of the new cultivars was multiple disease resistance, whereas, the gain in forage yield improvement was as low as 0.1–0.2% per year.

The current study was designed to examine the variability of morphological traits and yield of alfalfa cultivars, determine their disease resistance and select the most promising accessions for future breeding programmes.

2   Materials and Methods

Research was conducted in 2014 and 2015 at Institute of Agriculture, Research Centre for Agriculture and Forestry in a field of a six-course crop rotation. The soil of the experimental site is *Endocalcari–Endohypogleyic Cambisol Mg-n-w-can* (pH – 7.2–7.3, P_2O_5 – 201–270 mg kg^{-1} and K_2O – 101–175 mg kg^{-1}, humus – 2.0–2.46%). The nursery of alfalfa cultivars was maintained under natural infection pressure. Each accession was sown at a rate of 0.2 g scarified seed per 1 m in two 5-m long rows in three replications with a special hand-operated sowing machine "Plomatic 1R", (Wintersteiger, Austria). The distance between the rows was 0.5 m and between accessions – 1.0 m. The experimental material included accessions of alfalfa of different geographical origin (Table 1).

Table 1. Species, number and origin of alfalfa cultivars tested

Species of alfalfa	Number of cultivars	Country of origin	Species of alfalfa	Number of cultivars	Country of origin
M. sativa L.	2	Lithuania	*M. sativa* L.	2	Ukraine
M. varia M.	2	Lithuania	*M. sativa* L.	3	Slovakia
M. varia M.	1	Latvia	*M. sativa* L.	1	Netherlands
M. sativa L.	9	Romania	*M. varia* M.	3	Estonia
M. sativa L.	5	France	*M. sativa* L.	1	Poland
M. sativa L.	6	Czech Republic	*M. sativa* L.	1	Italy

Plant diseases and morphological traits were evaluated in 2014 and 2015. Percentage of spring black stem and leaf spot (SBSLS) and downy mildew (DM) severity was evaluated during the entire growing season, (where 1 score – 0%; 9 – ≥80%). Resistance to sclerotinia crown and stem rot (SCSR) was assessed on a 1–9 score scale two weeks after spring regrowth (where 1.0 – the most resistance; 9.0 – plant dead).

Morphological traits were measured on a 1–9 score scale (9 – being the best value). The seed yield (kg ha^{-1}), fresh and dry matter yields (t ha^{-1}) were measured. Plant height was measured after spring regrowth and at flowering stage (cm). The area under the disease progress curve (AUDPC) was calculated as the total area under the graph of SBSLS and DM severity against time, from the first scoring to the last.

$$\text{AUDPC} = \sum_{i=1}^{n-1} \left[(t_{i+1} - t_i)(y_i + y_{i+1})/2 \right];$$

Where "t" is time in days of each reading, "y" is the percentage of affected foliage at each reading and "n" is the number of readings (Campbell and Madden 1990).

The experimental data were statistically processed using analysis of variance (ANOVA). The cluster analysis was performed using Statistica 7.0.

3 Results and Discussion

Highly significant differences were observed for all traits. This considerable variability provides a good chance of improvement in the studied alfalfa genotypes. The highest coefficient of variability was recorded for seed yield (46.5%) (Table 2). Bodzon (2004) determined about 60% of the seed yield variability among alfalfa cultivars. Also, high variability coefficients was determined for resistance to Sclerotinia crown and stem rot (21.1%) and downy mildew (16.9%), fresh matter yield (11.7%), dry matter yield (14.3%) and plant height during spring regrowth (10.9%). The coefficients of variation of these traits indicate a high breeding potential of the genotypes.

Table 2. Analysis of morphological traits of 37 alfalfa cultivars over the two growing seasons 2014 and 2015

Trait	Average	CV %
SCSR (score)	3.0	21.1
DM *(AUDPC value)*	1221.6	16.9
BSLS *(AUDPC value)*	1813.0	5.9
SY (kg^{-1})	239.7	46.5
FMY (t ha^{-1}) per year	57.5	11.7
DMY (t ha^{-1}) per year	13.3	14.3
PH1 (cm)	29.1	10.9
PH2 (cm)	103.5	6.5
SD (score)	7.7	8.3

SCSR – sclerotinia crown and stem rot; *DM* – downy mildew; *BSLS* – black stem leaf spot; *SY* – seed yield; *FMY* – fresh matter yield; *DMY* – dry matter yield; *PH1* – plant height during spring regrowth; *PH2* – plant height at flowering stage; *SD* – stem density

Cluster analysis for 9 agro-morphological traits was carried out to measure the genetic distance among the 37 alfalfa cultivars tested (Fig. 1). Touil et al. (2009) divided 35 Mediterranean populations of cultivated alfalfa based on 9 morphological traits into 3 groups which differed significantly in plant height, yield, stem dry matter yield and other traits. All the 37 cultivars could be classified into four clusters (Fig. 1). Cluster 1 included four Lithuanian cultivars 'Antanė' (*M. sativa* L.), 'Birutė' (*M. varia* Marth.), 'Malvina' (*M. sativa* L.), 'Žydrūnė' (*M. varia* Marth.) and one Latvian cultivar 'Skriveru'. These cultivars exhibited comparable resistance to diseases. Cluster 2 included the largest group of 28 cultivars. In this group, the average values of the traits: resistance to downy mildew and black stem leaf spots, fresh and dry matter yield were found to be above the total mean. In this cluster, the cultivar 'Vali' (SK) stood out by the highest seed yield, fresh and dry matter yield, plant height during regrowth in spring and at flowering stage. Cluster 3 included only cultivar 'Palava' (CZ). Average values of its traits: resistant to black stem leaf spot, seed yield, fresh matter and dry matter yield were below the total mean. Cluster 4 included 3 Estonian cultivars 'Karluu', 'Juurlu' and 'Jõgeva 118'. In this group, the average values for traits: resistance to downy mildew, plant height during regrowth in spring and at flowering stage were below the total mean. The cultivars of this cluster differed in seed yield, average values of this cluster were below the mean of cluster 1. Cluster 4 exhibited the best stem density.

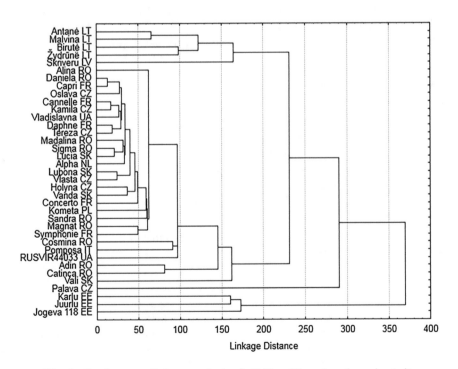

Fig. 1. Dendrogram of cluster analysis of alfalfa cultivars based on nine traits

4   Conclusions

This study revealed significant differences in the 9 traits tested among the 37 alfalfa cultivars investigated in the current study. Of all the tested cultivars, Lithuanian, Estonian and Latvian ones were the least damaged by downy mildew. The highest seed yield was produced by the cultivars 'Birutė', 'Žydrūnė', 'Skriveru', 'Karluu' and 'Jõgeva 118'. However, these cultivars were comparable in terms of fresh and dry matter yield, stem density, as well as resistance to black stem and leaf spot. The best performing genotypes, particularly those with the highest genetic distances, will be included in the next step of the alfalfa breeding process. These genotypes will be inter-crossed to form basic breeding germplasm for creating novel alfalfa cultivars.

Acknowledgments. This study was funded through the long-term research program "Genetics and purposeful change of genotypes of agricultural and forest plants" implemented by Lithuanian Research Centre for Agriculture and Forestry.

References

Anonymous (2012) Fodder plants. *Medicago sativa* L. Off J Eur Union 31:132–140. Common catalogue of varieties of agricultural plant species

Bodzon Z (2004) Correlations and heritability of the characters determining the seed yield of the long-raceme alfalfa (*Medicago sativa* L.). J Appl Genet 45:49–59

Bouton JH (2012) Breeding Lucerne for persistence. Crop Pasture Sci 63:95–106

Campbell CL, Madden LV (1990) Introduction to plant disease epidemiology. Wiley, New York

Castell-Miller CV, Zeyern RJ, Samac DA (2007) Infection and development of *Phoma medicaginis* on moderately resistant and susceptible alfalfa genotypes. Can J Plant Pathol 29:290–298

Ellwood SR, Kamphuis LG, Oliver RP (2006) Infection of sources of resistance to *Phoma medicaginis* isolates in *Medicago truncatula* SARDI core collection accessions, and multigene differentiation of isolates. Phytopathology 96:1330–1336

Herrmann D, Flajoulot S, Julier B (2010) Sample size for diversity studies in tetraploid alfalfa (*Medicago sativa*) based on co-dominantly coded SSR markers. Euphytica 171:441–446

Kanbe M, Mizuikami Y, Fujimoto F (2002) Improvement of resistance to *Sclerotinia* crown and stem rot of alfalfa through phenotypic recurrent selection. Jpn Agr Res Q 36:1–5

Lamb JFS, Sheaffer CC, Rhodes LH, Sulc RM, Undersander DJ, Brummer EC (2006) Five decades of alfalfa cultivars improvement: impact on forage yield, persistence, and nutritive value. Crop Sci 46:902–906

Liatukienė A, Žilvinas L (2010) Lucerne complex resistance to diseases. Scripta Hort Bot Univ Vyt Mag 14:98–104

Nagl N, Taski-Ajdukovic K, Barac G, Baburski A, Seccareccia I, Milic D, Katic S (2011) Estimation of genetic diversity in tetraploid alfalfa populations based on RAPD markers for breeding purpose. Int J Mol Sci 12:5449–5460

Rashidi M, Zang B, Cholami M (2009) Effect of different seedling rates on seed yield and same seed yield components of alfalfa (*Medicago sativa* L.). Int J Agric Biol 11:779–782

Sabanci CO, Ertus MM, Zorer Celebi S (2013) Collection, conservation and evaluation for forage yield of alfalfa landraces grown in East Anatolia. Turkish J Field Crops 18(1):46–51

Touil L, Guesmi F, Fares K, Zagrouba C, Ferchichi A (2009) Mineral composition and genetic variability of some Mediterranean populations of the cultivated alfalfa (*Medicago sativa* L.) supported by morphological markers. Asian J Plant Sci 8:1–10

Yaege JR, Stuteville DL (2000) Reaction in the annual *Medicago* core germplasm collection to two isolates of *Peronospora trifoliorum* from alfalfa. Plant Dis 84:521–524

Tetraploid Induction in *Lolium multiflorum*

O. Akinroluyo[✉], G. Statkevičiūtė, and V. Kemešytė

Institute of Agriculture, Lithuanian Research Centre for Agriculture and Forestry,
Kėdainiai reg., Lithuania
kelvin.akinroluyo@lzi.lt

Abstract. The induction of tetraploids from diploid cultivars of *L. multiflorum* is important in improving the germplasm. The aim of the study is to investigate the optimal concentration and time duration to induce tetraploids from diploid cultivars of *L. multiflorum*. Seedlings developing from excised embryos were subjected to different concentration of mitosis inhibitors and for different time intervals. The survival and induction percentages were determined for each induction approach.

Keywords: Annual ryegrass · Colchicine · Polyploidization

1 Introduction

Annual ryegrass (*Lolium multiflorum ssp. multiflorum*), also called Westerwolths ryegrass, is a leafy highly tillering grass of high palatability and digestibility. Annual ryegrass is also known to be one of the reliable cool-season grasses that have leaves that are rich in protein, vitamins and minerals in addition to being highly digestible and palatable to grazing animals (Humphreys et al. 2010).

Annual ryegrass occur naturally as diploids ($2n = 2x = 14$). However, the production of polypoid plants has been of interest to grass breeders to obtain differentiated genotypes to improve and maximize agronomic traits. Plants with different ploidy levels may differ in their growth habit, morphologically, physiological, cellular and biochemical aspects (Leitch and Leitch 2008).

Polyploid plants have been found to show a better resistance to both biotic and abiotic stress. In addition, polyploid plants have been found to display superior agronomic traits when compared to their diploids counterparts such as having a better tolerance to environmental stress (Comai 2005). These properties of polyploids enable them to be better adapted to a wider ecological range (Blanc and Wolfe 2004).

2 Materials and Methods

This study was performed at the Institute of Agriculture, Lithuanian Research Center for Agriculture and Forestry, Laboratory of Genetics and Physiology. Ten different cultivars of *L. multiflorum spp. multiflorum* were used (Table 1).

© Springer International Publishing AG, part of Springer Nature 2018
G. Brazauskas et al. (Eds.): *Breeding Grasses and Protein Crops in the Era of Genomics*, pp. 73–77, 2018.
https://doi.org/10.1007/978-3-319-89578-9_13

Table 1. List of cultivars used for polyploid induction.

Cultivar	Ploidy	Origin
Druva	2n	Latvia
Varpe	2n	Lithuania
Magloire	2n	France
Prompt	2n	France
Aramo	2n	France
Top speed	2n	France
Surrey nova	2n	USA
Weldra	2n	Netherlands
Grazer	2n	Germany
Shoot	2n	Denmark

The method described by Pašakinskienė (2000) was adopted for the sterilization of seeds and tetraploid induction. Seeds were surfaced sterilized and the embryos were excised from the seeds. Sterilized embryos were sprouted in a petri dish containing Gamborg B5 (Duchefa Biochemie) medium for 3–5days at a temperature of 24 °C. The coleoptiles were allowed to grow till up to 0.5 cm long and then transferred to a 4 °C refrigerator for 2 day to pause the process of mitosis. Prior to the treatment of plants with colchicine and amiprophos methyl (APM), the plants were transferred to growth chamber at 28 ° C for 1 h. The concentration and duration of the inhibitors are shown in Table 2.

Table 2. Treatment used for tetraploid induction in *Lolium multiflorum spp. multiflorum*

Mitosis inhibitors	Concentration	Duration
Colchicine	10 mM	3 or 4 h
Colchicine	8 mM	3 or 4 h
Amiprophos methyl	0.1 mM	4 h
Amiprophos methyl	0.05 mM	4 h
Amiprophos methyl	0.04 mM	4 h
Amiprophos methyl	0.03 mM	3 h
Amiprophos methyl	0.02 mM	3 h
Amiprophos methyl	0.015 mM	3 h

The ploidy levels of the survived plants were checked using a Partec PA flow cytometer. Also, a root tip squash technique for counting the chromosome number were used to verify the ploidy levels of 15 randomly selected plants

3   Results and Discussion

Chromosome doubling was achieved in a varying degree using different mitosis inhibitors however, the survival and induction rate depends on the affinity and toxicity of the inhibitors. In our experiments, different tetraploid induction rates were obtained using different concentrations and exposure times. Results from the Druva and Grazer cultivars of *L. multiflorum spp multiflorum* (Table 3) showed that both colchicine and APM were capable of inducing tetraploids from diploid cultivars. However, APM in higher concentrations appeared to be highly toxic to the plants. APM concentration of 0.05 mM with a 4 h exposure resulted in 12% survival rate of the treated plants but reducing the exposure time to 3 h and the concentration to 0.15 mM resulted in a 38.5% survival rate with a 40% induction rate in the Druva cultivar.

Table 3. Comparison of the survival and induction of tetraploid rate in 2 genotypes using amiprophos methyl and colchicine at different concentrations and exposure time

Mitosis inhibitors in various conc. time interval and temp.	Survival rate%		Induction rate%	
	Druva	Grazer	Druva	Grazer
10 mM colchicine, 4 h	24.5	28.8	71.4	61.4
8 mM colchicine, 3 h	59.7	52.5	25.6	37.1
10 mM colchicine, 3 h	40.7	63.3	68.1	61.4
8 mM colchicine, 4 h	41.2	32.9	42.5	55.8
0.1 mM amiprophos methyl, 4 h	0	0	0	0
0.05 mM amiprophos methyl, 4 h	12.5	14.9	57.1	50.0
0.04 mM amiprophos methyl, 4 h	21.7	24.0	30.8	45.8
0.03 mM amiprophos methyl, 3 h	17.0	18.9	37.5	44.4
0.02 mM amiprophos methyl, 3 h	22.2	23.2	30.0	31.8
0.015 mM amiprophos methyl, 3 h	38.5	–	40	–

The survival rate with colchicine and maximum tetraploid induction rates appeared to be better than with amiprophos methyl. A similar experiment to compare and evaluate the efficiency of colchicine and amiprophos methyl on double haploid production of onions was done by Foschi et al. (2013). Their findings revealed that colchicine was more efficient in doubling chromosomes than amiprophos methyl at the same exposure time although a higher concentration of colchicine was necessary to induce polyploidy than amiprophos methyl.

Weiler et al. (2014), found that treating seeds with mitosis inhibitors was more effective that seedling treatment of *Paspalum notatum*. They observed that treating seedlings with colchicine concentration of 0.1% and higher and for a longer duration (18–24 h) was highly toxic to the plants. Also, Pereira et al. (2014) showed that tetraploid induction in *Lolium multiflorum* using 15–20 day old seedlings was not possible at a concentration of 12.5 mM colchicine for 24 h. The optimal concentration resulted in 32% survival rate and 27% induction rate when the treatment is composed of 1% of DMSO in solution with 12.5 mM colchicine for 24 h (Pereira et al. 2014). However, our experiment showed that a high rate of tetraploid induction with lower concentrations of colchicine and a shorter exposure time was achievable. The optimal concentration for tetraploid induction in *Lolium multiflorum* (Grazer) is achieved using colchicine treatment with a concentration of 10 mM for 3 h which resulted in a 63.3% survival rate and a 61.4% induction rate.

Many factors such as the concentration of colchicine, plant genotype, the exposure time and the treated seedling organ have been to found to determine the efficiency of colchicine in inducing polyploids in *Rosa* species (Khosravi et al. 2008). Based on the results reported in Table 3, two combinations of colchicine concentration and exposure time were chosen to induce tetraploids in 8 different genotypes of *L. multiflorum*. Significant differences in the survival and induction rate among cultivars were observed.

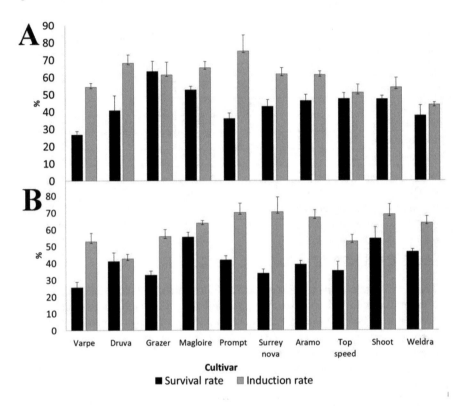

Fig. 1. Colchicine efficiency in inducing tetraploids in 10 diploid cultivars at different concentration and exposure time A (0.01 M colchicine, 3 h) B (0.008 M colchicine, 4 h). The error bar represents the standard error.

4   Conclusion

Induction of tetraploids from diploid cultivars of *L. multiflorum* can be achieved using both colchicine and APM. However, the efficiency of the mitosis inhibitors depends on the optimal concentration, exposure time, affinity and toxicity of the inhibitors. A concentration of 0.01 M and 0.008 M with an exposure time of 3 and 4 h respectively was found to be most efficient in inducing tetraploids from diploid cultivars of *L. multi-florum*.

Acknowledgements. The research project was funded by the Research Council of Lithuania, grant No. MIP-064/2015 (ADAPTGENAS).

References

Blanc G, Wolfe H (2004) Functional divergence of duplicated genes formed by polyploidy during Arabidopsis evolution. Plant Cell 16:1679–1691

Comai L (2005) The advantages and disadvantages of being polyploid. Nat Rev Genet 6:836–846

Foschi M, Martínez L, Ponce M, Galmarini C, Bohanec B (2013) Effect of colchicine and amiprophos-methyl on the production of in vitro doubled haploid onion plants and correlation assessment between ploidy level and stomatal size. FCA UNCUYO. 45(2):155–164

Humphreys M, Feuerstein U, Vandewalle M, Baert J (2010) Ryegrasses. In: Boller B, Posselt UK, Veronesi F (eds) Handbook of plant breeding: fodder crops and amenity grasses. Springer, New York, pp 211–260

Khosravi P, Kermani M, Nematzadeh G, Bihamta, Yokoya K (2008) Role of mitotic inhibitors and genotype on chromosome doubling of Rosa. Euphytica 160:267–275

Leitch A, Leitch I (2008) Genomic plasticity and the diversity of polyploid plants. Science 320:481–483

Pašakinskiene I (2000) Culture of embryos and shoot tips for chromosome doubling in Lolium perenne and sterile hybrids between Lolium and Festuca. Plant Breeding 119:185–187

Pereira R, Ferreira M, Davide L, Pasqual M, Mittelmann A, Techio V (2014) Chromosome duplication in *Lolium multiflorum Lam.* Crop Breed Appl Biotechnol 14:251–255

Weiler R, Krycki K, Guerra D, Simioni C, Dall'Agnol M (2014) Chromosome doubling in *Paspalum notatum* var. *saure* (cultivar Pensacola). Crop Breed Appl Biotechnol 15:106–111

The Sense of Testing Sheep Preference in Tall Fescue Variety Evaluation

M. Cougnon[1(✉)], J. Baert[2], and D. Reheul[1]

[1] Department of Plant Production, Ghent University, Ghent, Belgium
mathias.cougnon@ugent.be
[2] ILVO Plant, Melle, Belgium

Abstract. Tall fescue (*Festuca arundinacea* Schreb.) has a low digestibility and a low intake when grazed compared to ryegrasses. To make progress in these traits, we introduced a new way of testing candidate varieties (candivars) in our breeding program. We ran simultaneously a yield trial, in which yield and digestibility were measured, and a preference trial where sheep preference was evaluated using two methods: a qualitative (visual) and a quantitative measurement. In this paper we question whether such a preference test has an added value. Based on results of 18 candivars, we conclude that measuring sheep preference makes sense, because not all varieties with a high digestibility have good preference.

Keywords: *Festuca arundinacea* · Palatability · Digestibility

1 Introduction

In the light of adapting forage production to climate change, tall fescue (*Festuca arundinacea* Schreb.) is a grass species that gains importance in NW-Europe (Reheul et al. 2016). An important disadvantage of this species is however its low feeding quality. Compared to *Lolium sp.* both the digestibility of the forage as well as the voluntary intake of animals fed with tall fescue is lower (Cougnon et al. 2013).

In the past, breeders were successful in developing tall fescue varieties with improved digestibility (Johnston and McAneney 1994) or improved animal preference (Emile et al. 1992). However, the low emphasis on feeding value in official variety testing resulted in low breeding incentives for its further improvement. Recently, there is a consensus among European breeders and variety testing authorities, that official variety testing should pay more attention to parameters reflecting feed quality than to yield (Baert and Muylle 2016), particularly for species characterized by a low digestibility (e.g. tall fescue and cocksfoot (*Dactylis glomerata*)).

Since 2008, the Department of Plant Production of Ghent University runs a tall fescue breeding and research program. One of the aims of this program is to understand the factors that determine feeding quality of tall fescue and to breed varieties with an improved feeding quality adapted to NW European conditions (Cougnon 2013; Shahidi et al. 2016). When testing candivars, we are not only measuring yield and digestibility but we also monitor sheep preference. Several studies have shown that there is a strong positive correlation between the digestibility and animal preference in tall fescue

© Springer International Publishing AG, part of Springer Nature 2018
G. Brazauskas et al. (Eds.): *Breeding Grasses and Protein Crops in the Era of Genomics*, pp. 78–82, 2018.
https://doi.org/10.1007/978-3-319-89578-9_14

(Buckner and Burrus 1962; Gillet and Noël 1983; Shahidi et al. 2016). Hence, one can question the relevance of the effort to test both digestibility and sheep preference. The aim of this paper is to assess whether the co-testing of sheep preference with digestibility offers an added value for the evaluation protocol of tall fescue varieties.

2 Material and Methods

2.1 Plant material and Phenotyping

In April 2014 we established two trials with the same 18 candivars from our breeding programme and three reference varieties (*Barolex, Callina* and *Jugurta*). Annual fertilization in both trials was 300 kg N ha^{-1}, 60 kg P_2O_5 ha^{-1} and 350 kg K_2O ha^{-1}.

A first trial (called "**yield trial**" hereafter) was sown at the Flemish Institute of Agriculture and Fisheries Research (ILVO). Plots of 7.8 m^2 were sown in a randomized complete block design with three replicates at a density of 10000 viable seeds per plot. In 2015, the trial was harvested 4 times (28th April, 10th June, 22nd July, 28th September) using a plot harvester (Haldrup, Logstor, Denmark), weighing the fresh grass. A random subsample of the harvested material was taken from each plot, dried for at least 24 h at 70 °C allowing calculation of the dry matter yield (DMY). Afterwards, samples were ground and analysed for the digestibility of the organic matter (DOM) and the water soluble carbohydrates (WSC) concentration with NIRS; calibrated according to Tilley and Terry (1963) for DOM and Wiseman et al. (1960) for WSC.

A second trial (called "**preference trial**" hereafter) was established on the experimental farm of Ghent University in Melle. Candivars and reference varieties were sown in single rows of 10 m long in three replicates. Distance between rows was 50 cm. Sowing density was 300 viable seeds/m. At least two days before harvesting the yield trial, half of each row was fenced to prevent grazing, and the remainder of the trial was stocked with sheep (3 ewes and their lambs). Twice a day grazing preference was scored visually on a scale from 0 (no grazing at all), 1 (between 0% and 10% of standing biomass eaten), 2 (between 10 and 20% of standing biomass eaten).... to 9 (between 80 and 90% of standing biomass eaten). The stocking period ended as soon as one variety reached a score of 9. At that moment the proportion of the grass that had been grazed was quantified. Using a hedge trimmer, we harvested 3 m both of the grazed and ungrazed halfs of the rows at a cutting height of 5 cm. Harvested biomass was weighed and dried. This, therefore, resulted in two estimates of sheep preference: a visual (qualitative) and a quantitative estimate. The results of the first cut, are not reported, since results of this spring cut were biased due to differences in heading dates of the varieties.

2.2 Data Analyses

Statistical analyses were performed using the software package R. (R core team 2013). The relationship between different measured parameters was studied using a general linear model including the grazing period as covariate (ancova analysis).

3 Results and Discussion

There was no consistent significant regression between sheep preference, expressed both as the quantified proportion of grazed grass or as a visual score, and DOM nor between preference and WSC (Table 1). The absence of a significant relationship in all cuts between preference and DOM can be explained by the fact that not all varieties with a high DOM had a high preference: e.g. candivar 9 was among the candivars with the lowest preference in the three cuts, but in cut 2 and 4 it was among the three best for DOM (Fig. 1). To the contrary, varieties with a high preference generally had a good DOM: e.g. candivars 2, 3, 5. In former research (Shahidi et al. 2016), we found a significant, positive effect of DOM and WSC on sheep preference. However, this research was conducted with a selection of 17 genotypes with very different contrasting morphogenetic traits. Therefore, although there was a partial overlap in the information provided by the sheep preference and the analyzed digestibility, we conclude that sheep preference gave extra information: e.g. despite its good DOM, we will not withhold candivar 9 for further development based on its low sheep preference.

Table 1. Correlation coefficients between a qualitative or a quantitative estimate of sheep preference and digestibility of the organic matter (DOM) or the concentration of water soluble carbohydrates WSC. NS = non-significant; * = p-value between 0.05 and 0.01.

		Quantitative	Qualitative
DOM	Cut 1	0.35^{NS}	0.34^{NS}
	Cut 2	0.24^{NS}	0.30^{NS}
	Cut 3	0.46^{*}	0.63^{*}
WSC	Cut 1	0.16^{NS}	0.17^{NS}
	Cut 2	0.39^{NS}	0.42^{NS}
	Cut 3	0.21^{NS}	0.26^{NS}

As farm animals rarely can choose the grass variety they are offered, one can argue that there is no sense in putting effort to breed forage grass with a high preference (Falkner and Casler 1998). However, there are indications that varieties bred for higher preference resulted in higher intake and milk production (Emile et al. 1992). With this in mind, we are convinced that testing both digestibility and preference of candidate varieties is useful.

In correspondence with earlier work of Buckner and Burrus (1962), there was a very good correspondence between the quantitative estimation of the proportion grazed grass and the visual method ($R^2 = 0,84$) (Fig. 2, left). Taken all the results of the combined trials, the discriminative power of the visual score was high enough for reliable elimination of material with the worst digestibility.

There was a negative correlation between digestibility and dry matter yield of the studied cuts (p = 0.0012, $R^2 = 0.45$) (Fig. 2, right), indicating that an intensive selection for improved digestibility eventually may lead to a loss of yield. Breeding has to pay attention to combine an improved feeding value with a high yield.

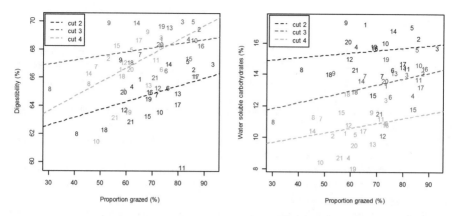

Fig. 1. Relationship between sheep preference (quantified as the proportion grazed grass) and digestibility (left) or the concentration of water soluble carbohydrates (right) for 18 candidate varieties and three reference varieties (19, 20, 21) of tall fescue in three cuts in 2015

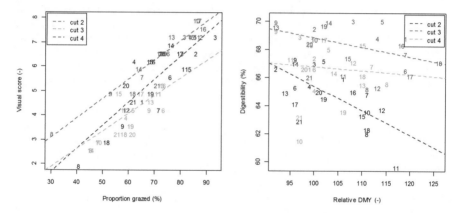

Fig. 2. Left: relationship between a visual score for grazing intensity and the quantified proportion of grazed grass. Right: relationship between digestibility and relative dry matter yield (DMY); 100 = average DMY of reference varieties 19, 20, 21.

4 Conclusion

Evaluating both sheep preference and digestibility of candivars is useful: not all varieties with a high digestibility have high preference. Using this strategy is a cost effective strategy to improve both digestibility and intake of newly developed varieties.

Acknowledgments. We are grateful to the students Adrien De Mullot de Villenaut and Shima Yazdani Foshtomi for their help in the gathering of the data.

References

Baert J, Muylle H (2016) Feeding value evaluation in grass and legume breeding and variety testing: Report of a debate. In: Roldan-Ruiz I, Baert J, Reheul D (eds) Breeding in a world of scarcity, proceedings of the 2015 meeting of the Eucarpia section "Forage Crops and Amenity Grasses". Springer, pp 307–311

Buckner R, Burrus P (1962) Comparison of techniques for evaluating palatability differences among tall fescue strains. Crop Sci 2:55–57

Cougnon M (2013) Potential in mixed swards and breeding of tall fescue (*Festuca arundinacea*). Doctoral dissertation, Ghent University, p 257

Emile JC, Gillet M, Ghesquière M, Charrier X (1992) Pâturage continu de fétuques élevées par des vaches laitières: amélioration de la production par l'utilisation d'une variété sélectionnée pour l'appétibilité. Fourrages 130:159–169

Falkner L, Casler M (1998) Preference for smooth bromegrass clones is affected by divergent selection for nutritive value. Crop Sci 38:690–695

Gillet M, Noël C (1983) La cafétéria d'auges, méthode d'étude de l'appétibilité des fourrages. Agronomie 3:817–822

Johnston DT, McAneney DMP (1994) Breeding for improved dry matter digestibility in tall fescue (*Festuca arundinacea*). Breeding for quality, proceedings of the 1994 meeting of the Eucarpia Fodder Crops section. Rijkstation voor Plantenveredeling, Merelbeke, pp 213–224

Reheul D and 25 others (2016) Report of the breeding debate. In: Roldan-Ruiz I, Baert J, Reheul D (eds) Breeding in a world of scarcity, proceedings of the 2015 meeting of the Eucarpia section "Forage Crops and Amenity Grasses". Springer, pp 313–319

R Core Team (2013) R: A language and environment for statistical computing. R Foundation for Statistical Computing, Vienna, Austria. http://www.R-project.org/

Shahidi R, Cougnon M, Struyf E, Van Waes C;, Van Labeke MC, Reheu D (2016) Parameters influencing preference by sheep in soft leaved tall fescue genotypes. In: Roldan-Ruiz I, Baert J, Reheul D (eds) Breeding in a world of scarcity, proceedings of the 2015 meeting of the Eucarpia section "Forage Crops and Amenity Grasses". Springer, pp 283–288

Tilley J, Terry R (1963) A two-stage technique for the in vitro digestion of forage crops. Grass Forage Sci 18(2):104–111

Wiseman HG, Mallack JC, Jacobson WC (1960) Determination of sugars in silages and forages. J Agric Food Chem 8:78–80

Chemical Composition of Zigzag Clover
(*Trifolium medium* L.)

B. Butkutė[1(✉)], N. Lemežienė[1], A. Padarauskas[2], E. Norkevičienė[1],
and L. Taujenis[2]

[1] Institute of Agriculture, Lithuanian Research Centre for Agriculture
and Forestry, Kėdainiai Reg., Lithuania
bronislava.butkute@lammc.lt
[2] Department of Analytical and Environmental Chemistry, Vilnius University,
Vilnius, Lithuania

Abstract. The study deals with the assessment of zigzag clover (*Trifolium medium* L.) for nutritional and mineral composition as well as for the distribution of isoflavones in whole aerial plant part at vegetative and full flowering stages, and morphological parts of flowering plants. Zigzag clover proved to be valuable protein- mineral- and isoflavone-rich legume crop with low NDF content. Plants of vegetative stage exhibited higher concentrations of most components tested. According to the total concentration of isoflavones, the plant parts ranked as follows: leaves ≫ stems ≫ flowers. The peculiarities of distribution of nutritive, mineral components and isoflavones, revealed in the study, could be useful for exploiting species both in animal nutrition and in the development of health-benefiting food ingredients.

Keywords: *Trifolium medium* · Mineral and nutritional composition
Isoflavones · Growth stage · Plant part

1 Introduction

As crop improvement through breeding critically depends on crop genetic resources, knowledge of the potential of natural populations is an essential starting point (Kizeková et al. 2013). Quality evaluation of wild species improves a basis of information on their specific characteristics. Herbaceous perennial legumes traditionally are used as highly nutritious fodder crops. Forage legumes can be classified as "protein crops" in the European Union (Phelan et al. 2015). The perennial legumes are rich in minerals, especially at vegetative growth stage (Butkutė et al. 2016). Isoflavones are one of the important groups of phenolic compounds in legumes and, like any bioactive compound they can have both positive and negative effects on animal and human health (Bennetau-Pelissero 2016). Moorby et al. (2004) found that lambs grazing red clover containing high concentrations of formononetin had slightly higher daily weight gains than those grazing low-isoflavone clover or perennial ryegrass. Phytoestrogens from forages fed to dairy cows increase concentrations of isoflavones and especially nonsteroidal estrogen equol in milk (Adler et al. 2015). The negative effect of phytoestrogens is primarily associated with deleterious influence on reproduction and

© Springer International Publishing AG, part of Springer Nature 2018
G. Brazauskas et al. (Eds.): *Breeding Grasses and Protein Crops in the Era of Genomics*, pp. 83–87, 2018.
https://doi.org/10.1007/978-3-319-89578-9_15

estrogen-dependent diseases (Bennetau-Pelissero 2016). Phytoestrogen values exceeding 1% of dry matter are associated with reproductive disorders in animals (Hloucalová et al. 2016). Therefore, it is not recommended to feed animals on diets rich in isoflavones during mating season and early pregnancy (Wocławek-Potocka et al. 2013).

For *T. medium* the data on nutritive quality and phytochemicals in plants and morphological fractions are extremely rare. The aim of this study was to assess nutritional potential and quantify isoflavones in zigzag clover, which naturally grow under a cool temperate climate environment, and to investigate isoflavone distribution in plants as influenced by plant part and growth stage.

2   Material and Methods

2.1   Experimental Design and Sample Preparation

Seeds of wild ecotype of *T. medium* L. (Catalogue No. at the Institute of Agriculture, Lithuanian Research Centre for Agriculture and Forestry ŽI-2148) were collected in natural habitat in Lithuania and the following year were sown in the germplasm collection of the perennial legumes spp. in single rows, 2.5 m long and 0.5 m apart. Plots were set up in the experimental site of the Institute of Agriculture (55°23′49″N; 23°51′40″E). The trials were laid out in a randomized complete block design, with four replications. Herbicides and fertilizers were not applied in trial. For chemical assays, composite samples from four field replications were collected at full flowering and vegetative stages. Aerial part of zigzag clover sampled at full flowering was divided into two subsamples. One of them was investigated as a sample of whole aerial plant part and the other was fractionated into morphological fractions: stems, leaves and flowers. The samples were chopped, oven-dried at 65(±5) °C and ground to pass a 1-mm screen using a cyclone mill. Chemical analyses were carried out in triplicate.

2.2   Analyses of Nutritional, Mineral and Isoflavone Composition

Crude protein was determined by the Kjeldahl method; crude fat – gravimetrically by the continuous Soxhlet extraction with hexane. NDF extraction was done on an ANKOM220 Fiber Analyzer (ANKOM Technology, Macedon, NY, USA, ANKOM method 08-16-06). Crude ash content was determined as the mass left after sample incineration at 550(±10) °C. Concentrations of soluble sugars were measured spectrophotometrically (M107, Camspec, UK) using the anthrone reagent (Zhao et al. 2010). Starch was determined in plant material residue after soluble sugar washing and hydrolysis with enzymes α-amylase and amyloglucosidase; released glucose was assayed following procedures described by Zhao et al. (2010). Potassium, calcium, magnesium, zinc and iron were quantified after digestion with the nitric acid plus hydrogen peroxide by flame atomic absorption spectroscopy using a Perkin Elmer model AAnalyst 200 (USA). Phosphorous was determined spectrophotometrically (Cary 50, Varian, USA) after sulfuric acid digestion of the samples and reaction with molybdate-vanadate.

Extraction, chromatographic separations and identification of four isoflavones (daidzein, genistein, formononetin and biochanin A) were performed according to our recently published procedure (Lemežienė et al. 2015). A Waters Acquity ultra performance liquid chromatography (UPLC) system (Waters, Milford, MA) equipped with a diode array (DAD) detector was used for analysis. Analytes in the extracts were identified by comparing the retention times with those of the corresponding standards.

Results were reported as the means ± standard deviations (SD) of three replicates of laboratory tests and were expressed on dry matter basis.

3 Results and Discussion

At full flowering, crude protein and ash concentrations in *T. medium* plants were relatively high and amounted to 167 and 97.7 g kg^{-1}, respectively (Table 1). Flowering plants contained 67.9 g kg^{-1} of readily digestible soluble sugars, on a par with starch of 20.7 g kg^{-1}. As expected, the nutritive quality of young plants was better than that of mature ones: NDF content of the plants at vegetative growth stage was considerably lower (309 g kg^{-1}) than for ones at full flowering (359 g kg^{-1}). Leafy young plants had a much higher proportion of crude protein (190 g kg^{-1}), starch (39.7 g kg^{-1}), ash (104 g kg^{-1}) and sugars (79.0 g kg^{-1}) than fully flowering plants. Young zigzag clover was richer in calcium, zinc and iron than fully-flowering plants, whereas the concentrations of potassium, magnesium and phosphorus were of the same level for fully flowering and young plants. Overall results on nutritional and mineral composition proved that zigzag clover could to be classified as "protein crop" with high mineral and low NDF contents. Differences in nutritional composition between young and flowering plants arose from several impacts: dilution effect (decrease of concentrations of crude protein, ash, and fat in herbage yield, which is increasing with advancing plant maturity), change of leaf to stem ratio during plant development, and environmental conditions.

Species *T. medium* had very high concentration of isoflavones (Table 2). The total amount of compounds was estimated of 24.45 mg g^{-1} at full flowering stage and of 28.66 mg g^{-1} at vegetative stage. Despite the great qualitative and quantitative variation among samples of whole aerial plant parts of two growth stages and individual morphological fractions, formononetin was a dominant isoflavone (5.28–15.07 mg g^{-1}) and it was accompanied by fairly high concentrations of biochanin A (2.10–14.67 mg g^{-1}). At an early growth stage, formononetin, biochanin A and genistein were found in significantly greater concentrations than at full flowering stage. Daidzein concentration in young plants (0.149 mg g^{-1}) was considerably lower than that in flowering plants (0.235 mg g^{-1}). Among the plant parts, the highest concentration of the sum isoflavones was determined in the leaves (31.81 mg g^{-1}) compared with stems (15.58 mg g^{-1}) and flowers (7.77 mg g^{-1}). In contrast to other three isoflavones, the daidzein content in stems was higher than in leaves.

Our results on isoflavone distribution depending on growth stage and aerial parts were in agreement with the literature data concerning red clover (Booth et al. 2006; Saviranta et al. 2008). Reliable data on isoflavones quantification in *T. medium* plants are currently not available. It should be noted that the concentrations of phytoestrogens investigated in zigzag clover plants are several times as high as those in other legumes

Table 1. The proximate and mineral composition of zigzag clover (mean ± SD).

Sample characteristic	Protein	Fat	NDF	Soluble sugars	Starch	Ash
	g kg⁻¹ DM					
Full flowering stage	167 ± 8.3	41.2 ± 2.2	359 ± 17.9	67.9 ± 3.7	20.7 ± 2.0	97.7 ± 2.5
Vegetative stage	190 ± 7.8	42.6 ± 0.2	309 ± 12.0	79.0 ± 4.6	39.7 ± 2.3	104 ± 4.9
Sample characteristic	Potassium	Calcium	Magnesium	Phosphorus	Zinc	Iron
	g kg⁻¹ DM				mg kg⁻¹ DM	
Full flowering stage	22.7 ± 0.8	20.1 ± 0.9	5.41 ± 0.23	2.89 ± 0.06	25.9 ± 0.45	187 ± 7.2
Vegetative stage	22.1 ± 0.2	22.2 ± 1.1	5.16 ± 0.11	2.78 ± 0.23	29.9 ± 0.18	230 ± 3.6

Table 2. Isoflavone distribution in zigzag clover plants of different growth stages and morphological parts of flowering plants.

Sample characteristic	Isoflavone concentration ± SD (mg g⁻¹)				
	Formononetin	Biochanin A	Daidzein	Genistein	Sum
Full flowering stage	12.11 ± 0.307	9.64 ± 0.209	0.235 ± 0.015	1.25 ± 0.049	23.24
Vegetative stage	14.49 ± 0.270	11.46 ± 0.289	0.149 ± 0.011	2.56 ± 0.121	28.66
Stems	10.81 ± 0.265	3.17 ± 0.099	0.300 ± 0.014	1.29 ± 0.047	15.58
Leaves	15.07 ± 0.371	14.67 ± 0.375	0.248 ± 0.014	1.81 ± 0.077	31.81
Flowers	5.28 ± 0.181	2.10 ± 0.083	0.0223 ± 0.003	0.363 ± 0.020	7.77

(soybean, red clover) well known and widely used as sources of phytoestrogens (Lemežienė et al. 2015; Seguin et al. 2004). Hence the *T. medium* plant material could become of particular interest as a source of phytoestrogens. Nevertheless, the main application of perennial legume species is forages for cattle and other livestock and due to the very high amount of isoflavones in *T. medium* plants, their use in forage during mating season and early pregnancy should be cautious. However, the negative effects can be avoided when plants are not grown in monoculture. Combination of forages differing in the composition of nutrients and plant secondary metabolites help animals maintain intake and increase nutrient utilization as well as improve production quality (Moorby et al. 2004; Adler et al. 2015).

4   Conclusion

Zigzag clover is valuable protein-, mineral- and isoflavone-rich perennial legume with low NDF content. The peculiarities of distribution of nutritive, mineral components and isoflavones, revealed in the study, could be useful for exploiting species both in animal nutrition and in the development of nutraceuticals.

Acknowledgments. This study was funded through long-term research programs "Biopotential and Quality of Plants for Multi-functional Use" and "Genetics and purposeful change of genotypes of agricultural and forest plants" implemented by Lithuanian Research Centre for Agriculture and Forestry.

References

Adler SA, Purup S, Hansen-Møller J, Thuen E, Steinshamn H (2015) Phytoestrogens and their metabolites in bulk-tank milk: effects of farm management and season. PLoS ONE 10: e0127187. https://doi.org/10.1371/journal.pone.0127187

Bennetau-Pelissero C (2016) Risks and benefits of phytoestrogens: where are we now? Curr Opin Clin Nutr Metab Care 19:477–483

Booth NL, Overk CR, Yao P, Totura S, Deng Y, Hedayat A, Bolton JL, Pauli GF, Farnsworth NR (2006) Seasonal variation of red clover (*Trifolium pratense* L., Fabaceae) isoflavones and estrogenic activity. J Agric Food Chem 54:1277–1282

Butkutė B, Lemežienė N, Dagilytė A, Cesevičienė J, Benetis R, Mikaliūnienė J, Rodovičius H (2016) Mineral element, total phenolic content and antioxidant capacity of the aerial plant parts and seeds of perennial legumes. Commun Soil Plant Anal 47(sup1):36–45

Hloucalová P, Skládanka J, Horký P, Klejdus B, Pelikán J, Knotová D (2016) Determination of phytoestrogen content in fresh-cut legume forage. Animals 6:43. https://doi.org/10.3390/ani6070043

Kizeková M, Martincová J, Jančová Ľ (2013) Evaluation of quantitative and qualitative traits of wild ecotypes of forage grasses. Agriculture 59(1):21–32

Lemežienė N, Padarauskas A, Butkutė B, Cesevičienė J, Taujenis L, Norkevičienė E, Mikaliūnienė J (2015) The concentration of isoflavones in red clover (*Trifolium pratense* L.) at flowering stage. Zemdirb Agric 102:443–448

Moorby JM, Fraser MD, Theobald VJ, Wood JD, Haresign W (2004) The effect of red clover formononetin concentration on live-weight gain, carcass characteristics and muscle equol concentration of finishing lambs. Anim Sci 79:303–313

Phelan P, Moloney AP, McGeough EJ, Humphreys J, Bertilsson J, O'Riordan EG, O'Kiely P (2015) Forage legumes for grazing and conserving in ruminant production systems. Crit Rev Plant Sci 34:281–326

Saviranta NMM, Anttonen MJ, Von Wright A, Karjalainen RO (2008) Red clover (*Trifolium pratense* L.) isoflavones: determination of concentrations by plant stage, flower colour, plant part and cultivar. J Sci Food Agric 88:125–132

Seguin P, Zheng W, Smith DL, Deng W (2004) Isoflavone content of soybean cultivars grown in Eastern Canada. J Sci Food Agric 84:1327–1332

Wocławek-Potocka I, Mannelli C, Boruszewska D, Kowalczyk-Zieba I, Waśniewski T, Skarżyński DJ (2013) Diverse effects of phytoestrogens on the reproductive performance: cow as a model. Int J Endocrinol. Article ID 650984

Zhao D, MacKown CT, Starks PJ, Kindiger BK (2010) Rapid analysis of nonstructural carbohydrate components in grass forage using microplate enzymatic assays. Crop Sci 50:1537–1545

Wild and Semi Natural Ecotypes of Perennial Grasses and Legumes – for Breeding Purposes

N. Lemežienė, V. Stukonis, V. Kemešytė, and E. Norkevičienė[(⊠)]

Institute of Agriculture, Lithuanian Research Centre for Agriculture and Forestry, Kėdainiai Reg., Lithuania
egle.norkeviciene@lammc.lt

Abstract. During 1994–2014 perennial grass and legumes wild and semi-natural ecotypes were collected in different geographic locations and regions of Lithuania, Baltic littoral (Latvia), Kaliningrad region (Russia), Precarpathian, Carpathian, Zacarpathian and Polese regions (Ukraine). Altogether about 2500 seed accessions of 32 species of grasses and 18 species of legumes were collected.

On the basis of most promising wild ecotypes, the following forage type varieties of grasses and legumes have been developed: smooth-stalked meadow grass (*Poa pratensis* L.) 'Rusnė', black medic (*Medicago lupulina* L.) 'Arka', meadow foxtail (*Alopecurus pratensis* L.) 'Valentas', redtop (*Agrostis gigantea* Roth.) 'Violeta', reed canary grass (*Phalaris arundinacea* L.) 'Pievys DS'. For the development of forage-type varieties of some species, such as timothy, ryegrass and meadow fescue, wild ecotypes were in most cases insufficiently luxuriant. About 3% of the forage-type plants of perennial grasses exhibited resistance to foliar diseases.

According to a complex of valuable agro morphological characteristics, small part (about 1.5%) of the wild ecotypes of various species were found to be promising for the application in turf grass breeding. Some turf grass varieties have been developed on the basis of the most valuable wild ecotypes: Canada bluegrass (*Poa compressa* L.) 'Odrė DS', wood bluegrass (*Poa nemoralis* L.) 'Luka DS', red fescue (*Festuca rubra* L.) 'Gludas', 'Varius' and 'Alkas', hard fescue (*Festuca trachyphylla (Hack.) Krajina*) 'Astravas'.

Keywords: Natural habitats · Forage type · Turf grass · Promising ecotypes

1 Introduction

Conservation of genetic resources of forage grasses and legumes in Lithuania was started in 1994 when plant breeders began activities in the Nordic–Baltic plant genetic resources (PGR) conservation project. The Nordic-Baltic PGR project was initiated by the Nordic Gene Bank. The main objective of the project was to develop the national PGR conservation network in Baltic countries (Thorn 2001; Weibull 2001).

Until 1994 natural ecotypes of forage grasses and legumes were collected only by initiative of individual plant breeders in a limited scale Only small quantity of most valuable wild ecotypes of a small number of species was involved in breeding pro-grammes. In most cases initial plant breeding material has been received from All–Union N.I. Vavilov Research Institute of Plant Industry, currently N. I. Vavilov Research

© Springer International Publishing AG, part of Springer Nature 2018
G. Brazauskas et al. (Eds.): *Breeding Grasses and Protein Crops in the Era of Genomics*, pp. 88–95, 2018.
https://doi.org/10.1007/978-3-319-89578-9_16

Institute of Plant Industry (Saint-Petersburg, Russia). Upon restoration of independence in Lithuania, the links with this institution were practically severed.

One of the most important tasks of the Global Plan of Action for Conservation and Sustainable Utilization of PGR for Food and Agriculture is establishing strong National programs (Gass et al. 1999). In 1994–1997 the Ministry of Education of the Republic of Lithuania funded the first research program "Genetic Resources of Cultivated Plants". Another research program "Investigation and conservation of genetic resources of cultivated plants and home animals" was approved and implemented in 1998–2002. In 2004–2008 the program "The Scientific Researches of Plant Genetic Resources" was conducted. Bilateral Lithuanian–Ukrainian program in the area of collecting and evaluating forage grasses and legumes, were implemented in 2007–2010 and in 2012–2013. These programs were also funded by the Ministry of Education and Science of the Republic of Lithuania (till 2010) later – by the Research Council of Lithuania.

Financial support for the PGR program was provided by Ministry of Agriculture and Ministry of Environment of Republic of Lithuania as well. Ministry of Environment of Republic of Lithuania still supports financially plant field collection maintenance and updating program.

Active participation in the program created suitable conditions for collection, investigation and preservation of PGR of forage grasses and legumes. Special attention during programs implementation period was paid to wild and semi natural ecotypes as a very valuable pre-breeding material. Wild and semi natural ecotypes are adapted to the local climatic and growing conditions because over thousands years of natural selection left only genes that encode the most necessary features and characteristics (Fjellheim et al. 2007; Tucak et al. 2009; Duller et al. 2010; Smýkal et al. 2014). Lithuania has still its natural habitats, however the fate of ecotypes is unclear because outside the state laws protected areas they can be irretrievably lost or destroyed. For this reason, it was urgent to collect local ecotypes in natural habitats of Lithuania and neighboring countries. Collected wild and semi-wild ecotypes should be investigated for their use in breeding.

The aim of this paper was to present the results of morpho-agronomical traits evaluation and ecotype breeding of perennial grass and legumes over the period 1996–2016.

2 Collecting of Wild and Semi Natural Ecotypes

On the basis of the international experience in the conservation of genetic resources (Tyler 1987) a list of the protected species, as well as the collecting principles of wild ecotypes (description form of natural habitats, size of samples, character of collecting, etc.) have been prepared. Inclusion of protected species in the priority list was based on certain conditions:

(1) Different trends of utilization. Species for forage, turf, biofuel, recultivation, green fertilizer, resowing after severe winter and other.
(2) Climate change. Drought resistance grass species become more important due to their origin from South Europe and Asian countries. These plant species have a

deep root system, which is adapted to assimilate water of deeper soil layers: *Medicago sativa* L., *Medicago* x *varia* Martyn, *Onobrychis viciifolia* Scop.

(3) Specific agricultural purposes. New collects of plant species which are able to grow in acid (*Lotus corniculatus* L.) or wet soil (*Phalaroides arundinacea* L.).

During 1994–2014 forage/turf grass and legumes wild and semi-natural ecotypes were collected in different geographic locations and regions of Lithuania, Baltic littoral (Latvia), Kaliningrad region (Russia), west and north part of Ukraine (Precarpathian, Carpathian, Zacarpathian and Polese regions) (Fig. 1).

Fig. 1. A map of collection areas.

Altogether about 2500 seed accessions of 32 species of grasses and 18 species of legumes were collected.

Passport data were prepared for each accession. Global Position System (GPS) was used to precisely identify the location of the habitat. An accession of cross pollinated perennial grass and legumes species was composed of seeds collected from dozen plants, but not less than 100 inflorescences from an ecologically homogeneous area of 100 to 1000 m^2 of each collecting site (Tyler et al. 1984; Yonezawa 1985).

Seeds were collected in more than 650 natural habitats: (1) on the slopes of hills and mountains (till 15% of all natural habitats), (2) on river and lake valleys natural meadows at the edge of forests that have not been touched by any human activity since time immemorial (about 80%), (3) on temporarily abandoned soils in formerly culti-vated meadows and pastures and stalls where human interference has been limited for the last 25–30 years (about 5%).

Lithuania and *Latvia* is a countries of flats. Situated between 56–53 latitude and 20–26 longitude (Lithuania) and 55–58 latitude and 21–28 longitude (Latvia). Most of the habitats were situated not higher than 10 m above sea level. The highest hills were only in 100–250 m above sea level.

Kaliningrad region (Russia) is located on the coast of the Baltic Sea. Situated between 55–54 latitude and 19–22 longitude. Most of the habitats were situated not higher than 10 m above sea level.

West and north part of Ukraine (Precarpathian, Carpathian, Zacarpathian and Polese regions). Situated between 47–51 latitude, 22–30 longitude. Most of the habitats were situated higher than 150 m above sea level. Carpatian mountains habitats were located in 500–1700 m above sea level.

3   Results of Ecotype Characterization/Evaluation

Characterization/evaluation trials of wild and semi-natural ecotypes of various forage and turf grass species were carried out at the Institute of Agriculture, Lithuanian Research Centre for Agriculture and Forestry during the period 1999–2014.

Forage and turf grasses. Most of the smooth – stalked meadow grass *Poa pratensis* L. wild ecotypes collected in the meadows and grasslands of riversides and strands and the outskirts of deciduous forest were characterised by luxuriance and corresponded rather to the forage type (Lemežienė and Kanapeckas 2002). Wild ecotypes collected on an infertile sandy soil (seacoast, hill tops, quarries) as well as intensively trodden places (churchyards, graveyards, bathing places, etc.) were shorter and less luxuriant, and had narrower leaves (lawn-type). For the development of forage-type varieties of some species, such as timothy, perennial ryegrass and meadow fescue, wild ecotypes were in most cases insufficiently luxuriant. About 3% of the forage-type plants of perennial grasses exhibited resistance to foliar diseases (Lemežiene and Kanapeckas 2008). According to a complex of valuable agro morphological characteristics, only about 1.5% of the wild ecotypes of various species will be promising for the application in turf grass breeding (Lemežiene and Kanapeckas 2008).

The eight narrow-leaved *Festuca* species – *F. filiformis* Pourr., *F. ovina* L., *F. pallens* Host.; *F. psammophila* (Hack. ex Čelak.) Fritsch., *F. pseudovina* Hack. ex Wiesb., *F. macutrensis* Zapał., *F. valesiaca* Schleich. ex Gaud., *F. rupicola* Heuff. are rare (except for *F. ovina*) in West Ukraine's Pre-Carpatian region and were found to grow only in specific habitats, i.e. hill and small mountain slopes, on sands and calcareous mountain rock exposures. In the genetic collection, according to such morphological traits as shortness, leaf narrowness and colour they belonged to turf-type grasses (Kanapeckas et al. 2010).

Agrostis capiliaris L. Seed accessions of 100 wild ecotypes, collected in Lithuania, Latvia and North Ukraine were assessed for agro biological traits: 1000 seed weight, plant height, panicle length, vegetative leaf length and width, beginning of inflorescence emergence etc. The average values of the tested traits did not significantly differ between the countries, except the date of inflorescence emergence – Ukrainian ecotypes distinguished by earliness. Rather high variation of agro-biological traits (CV = 11.0–57.7%) between ecotypes within species can be successfully used in turf grass breeding (Stukonis and Šlepetys 2013).

Poa pratensis L. Most of the 179 ecotypes (about 85%) were found to be of forage type, while the rest (about 15%) could be used for amenity purposes. Most of the ecotypes collected in Lithuania were sufficiently seedy. The highest seed productivity was demonstrated by the tall-growing smooth-stalked meadow grass. Lawn-type genotypes were often insufficiently seedy. Most of the smooth-stalked meadow grass accessions were not resistant to mildew, leaf spots and rust. Disease resistance was noted only in 4% of the genotypes (Lemeziene and Kanapeckas 2002).

Festuca pratensis Huds., *F. arundinacea* L. *and F. rubra* L. Wild ecotypes of various *Festuca* genera species, collected in the natural habitats of west Ukraine (Lvov, Ivano-Frankovsk, Ternopol and other areas) were tested. In terms of productivity, wild ecotypes of *F. pratensis and F. arundinacea* lag behind (*LSD05*) the standard Lithuanian varieties significantly. However, according to other traits (different development rate, disease resistance, plant height, etc.) of these ecotypes can be promising for breeding. All the *F. rubra* wild ecotypes were forage–type – tall–growing ones (Kanapeckas et al. 2010).

Lolium perenne L. The investigations of perennial ryegrass (120 accessions) show that the genetic diversity of the phenotypic traits depends on the origin of the ecotypes. High variation coefficient of plant height, flag leaf length and width, inflorescence length (CV% = 22.64–41.41) was detected among Ukrainian ecotypes (Statkeviciute et al. 2012). In the later studies the average variation coefficient of the same phenotypic traits (CV% = 10.57–16.39) was assessed among natural ecotypes from Lithuania, Ukraine, Latvia and Kaliningrad region (Russia) (Kemešytė et al. 2014, 2016). The lowest polymorphism was established for winter survival (Statkeviciute et al. Statkeviciute 2012; Kemešytė et al. 2014, 2016). The re-growth in spring and biomass were most stable traits and variation between ecotypes was CV = 22.69 and 22.38% (Statkeviciute et al. 2012; Kemešytė et al. 2014). Natural ryegrass ecotypes are used in the breeding programs as parental genotypes with the aim to create a new, more varied breeding material.

Wild narrow-leaved species of *Festuca* genus. Field experiments in Lithuania were designed to assess six narrow-leaved species of *Festuca* genus, for their suitability to be used as turf. The following *Festuca ovina* L. (*sensu lato*) species: *Festuca ovina* 22 populiations, *F. polesica* 3, *F. psammophila* 7, *F. trachyphylla* 21, *F. sabulosa* 3, *F. pseudovina* 1 were identified by morpho-anatomical methods. The following parameters were assessed in the variety trials: field germination speed, sward density, sward colour, sward green colour in late autumn, grow in spring, re–growth after cuts, leaf width and etc. According to the majority of the evaluated traits, wild populations of *F. pseudovina* Hackel ex Wiesb., *F. ovina* L. and *F. trachyphylla* (Hack.) Krajina species are the most suited for turf establishment. Other species, as *F. wolgensis* P. Smirnov, *F. psammophila* (Hack. ex Čelak.) Fritsdch. and *F. sabulosa* (Anddersson) H. Lindb. are less suited for turf establishment due to lower stand density and their susceptibility to rust in late autumn (Stukonis 2009; Stukonis et al. 2010). *Festuca ovina* is distributed all over Lithuania. It occurs most frequently and abundantly in pine forest zones and continental natural grasslands, especially in *Nardetea*. It is quite common in sandy communities and is less frequent in Central Lithuania's productive soil regions (Stukonis et al. 2009).

Trifolium repens L. White clover forms *hollandicum*, *giganteum* and *silvestre* differed insignificantly in some parameters of chemical composition. The wild ecotypes (*silvestre* form) had the lower crude protein content in average. The content of cyanogenic glucosides depended on the growth stage of plants, meteorological conditions of the growing season and time of harvesting. The herbage of the first cut wild ecotypes (36 accessions) contained smaller amount of cyanogenic glucosides (8.7–18.3) than the hybrid variety 'Sūduviai' (13.2–21.0 mg/100 g dry matter) (Sprainaitis and Paplauskiene 2002).

Trifolium spp. Clover has attracted considerable interest not only as valuable livestock forage but as an important source of isoflavones also. The current study was aimed to assess the variation of isoflavones concentration in clover species grown under a cool temperate climate environment in Lithuania. Isoflavone contents were quantified in the plant parts of 21 entries belonging to five perennial species of the genus *Trifolium (T. pratense, T. repens, T. medium, T. rubens and T. pannonicum)*. According to the averaged sum of the isoflavones, the 5 clover species ranked as follows: *T. medium* $(7.54–3.62–2.31$ mg g^{-1}) > *T. pratense* > *T. rubens* > *T. pannonicum* > *T. repens* $(0.191–0.204–0.171$ mg g$^{-1})$. With regard to isoflavone concentration and it variability within species, some accessions of *T. medium* and *T. pratense* can be considered a highly promising source of phytoestrogens (Vilčinskas 2010; Dabkevičienė et al. 2012; Butkute et al. 2014).

Trifolium pratense L. The objective of study was to determine the contents of isoflavones in Lithuanian diploid and tetraploid cultivars and wild ecotypes of red clover in aerial plant part and separately in stems, leaves and flowers at flowering stage. The highest concentration of isoflavones at flowering stage was accumulated in leaves. The average total concentration of all four isoflavones was as follows: 12.29 mg g^{-1} DM in leaves, 2.93 mg g^{-1} DM in stems and 1.42 mg g^{-1} DM in flowers. It was established that the relationship between isoflavone concentration and genotype ploidy, as well as the relationship between isoflavone concentration in cultivated varieties and wild ecotypes was not statistically significant (Lemežiene et al. 2015).

4 Results of Ecotype Breeding

The main method used in the development of turf grass varieties was individual selection. Experimental evidence suggests that most of the ecotypes did not possess valuable characteristics and biological properties for turf grass breeding. Tall-growing, forage-type plants with wide leaves and a thin bush prevailed. Only a small portion of local ecotypes were distinguished for the great diversity of the tested characteristics and traits.

On the basis of wild ecotype selection forage type varieties of grasses and legumes have been developed: smooth-stalked meadow grass (*Poa pratensis* L.) 'Rusnė', black medic (*Medicago lupulina* L.) 'Arka', meadow foxtail (*Alopecurus pratensis* L.) 'Valentas', redtop (*Agrostis gigantea* Roth.) 'Violeta', reed canary grass (*Phalaris arundinacea* L.) 'Pievys DS'.

Some turf grass varieties have been developed on the basis of the most promising wild ecotypes: Canada bluegrass (*Poa compressa* L.) 'Odrė DS', wood bluegrass (*Poa nemoralis* L.) 'Luka DS', red fescue (*Festuca rubra* L.) 'Gludas', 'Varius'and 'Alkas', hard fescue (*Festuca trachyphylla (Hack.) Krajina*) 'Astravas'.

5 Conclusions

About 3% of the forage-type plants of perennial grasses exhibited resistance to foliar diseases. Wild ecotypes of some species, such as timothy, perennial ryegrass and meadow fescue in most cases were insufficiently luxuriant for the direct development of forage-type varieties. About 1.5% of wild ecotypes of smooth-stalked meadow grass and red fescue species will be promising for the application in the forage and turf grass breeding. According to a complex of valuable agro morphological characteristics five varieties for forage and six – for turf have been developed. The relationship between isoflavone concentration in cultivated varieties and wild ecotypes was not statistically significant. Most valuable wild ecotypes were selected for further breeding works and storage in Plant Gene Bank.

Acknowledgments. This study as partly funded through the long-term research program "Genetics and purposeful change of genotypes of agricultural and forest plants" implemented by the Institute of Agriculture, Lithuanian Research Centre for Agriculture and Forestry.

References

Butkutė B, Lemežienė N, Dabkevičienė G, Jakštas V, Vilčinskas E, Janulis V (2014) Source of variation of isoflavone concentrations in perennial clover species. Pharm Mag 10(37):181–188

Dabkevičienė G, Butkute B, Jakštas V, Lemežienė N, Vilčinskas E, Janulis V (2012) Distribution of formononetin, daidzein and genisteinin *Trifolium* species and their aerial plant parts. Chemija 24(3):306–311

Duller S, Thorogood D, Bonos SA (2010) Breeding objectives in amenity grasses. In: Boller B, Posselt U, Veronesi F (eds) Fodder crops and amenity grasses. Handbook of plant breeding, vol 5. Springer, New York, pp 137–161

Fjellheim S, Blomlie AB, Marum P, Rognli OA (2007) Phenotypic variation in local population and cultivars of meadow fescue – potential for improving cultivars by utilizing wild germplasm. Plant Breed 126:279–286

Gass T, Frese L, Begemann F, Lipman E (comps.) (1999) Implementation of the Global Plan of Action in Europe – conservation and sustainable utilization of plant genetic resources for food and agriculture, 396 p

Kanapeckas J, Lemežienė N, Stukonis V, Kemešytė V, Konik GS (2010) Evaluation of morphological traits and biological characteristic of the wild ukrainian ecotypes of *Festuca* species. Vagos Moksl Darb 86(39):7–12 (in Lithuanian)

Kemešytė V, Statkevičiūtė G, Aleliūnas A, Brazauskas G (2016) Variation for plant architecture traits within a perennial ryegrass association mapping population. In: Roldán-Ruiz I, Baert J, Reheul D (eds) Breeding in a world of scarcity. Springer, Cham, pp 41–45

Kemešytė V, Statkevičiūtė G, Lemežienė N, Brazauskas G, Ivanciv R (2014) Morphological variation between perennial ryegrass (*Lolium perenne* L.) wild ecotypes. In: Sokolović D, Christian H, Radović J (eds) Quantitative traits breeding for multifunctional grasslands and turf. Springer, Dordrecht, pp 81–84

Lemežienė N, Kanapeckas J (2008) Selection efficiency of wild ecotypes for forage and turf grass breading. Biologija 54:125–128

Lemežienė N, Kanapeckas J (2002) Evaluation of smooth-stalked meadow grass wild ecotypes. Biologija 4:72–75

Lemežienė N, Padarauskas A, Butkutė B, Cesevičienė J, Taujenis L, Norkevičienė E, Mikaliūnienė J (2015) The concentration of isoflavones in red clover (*Trifolium pratense* L.) at flowering stage. Zemdirb Agric 102(4):443–448

Smýkal P, Coyne C, Ambrose M, Maxted N, Schaefer H, Blair M, Berger J, Greene S, Nelson M, Besharat N, Vymyslický T, Toker C, Saxena R, Roorkiwal M, Pandey M, Hu J, Li Y, Wang L, Guo Y, Qiu L, Redden R, Varshney R (2014) Legume crops phylogeny and genetic diversity for science and breeding. Crit Rev Plant Sci 34:43–104

Sprainaitis A, Paplauskiene V (2002) White clover breeding material and wild ecotypes for the formation of genetic collection. Biologija 4:15–18

Statkevičiūtė G, Kemešytė V, Lemežienė N, Konik GS, Chomiak M (2012) Genetic diversity of perennial ryegrass wild ecotypes based on agrobiological traits. Proc Latvian Acad Sci Sect B 66(4/5):168–171

Stukonis V, Lemežienė N, Kanapeckas J (2009) Distribution and polymorphism of seep fescue (*Festuca ovina* L.) in Lithuania. Vagos 81:68–74 (in Lithuanian)

Stukonis V, Lemežienė N, Kanapeckas J (2010) Suitability of narrow-leaved *Festuca* species for turf. Agron Res 8(3):729–734

Stukonis V, Šlepetys J (2013) The assessment of *Agrostis capiliaris* wild populations for use in turf grass breeding. Proc Latvian Acad Sci Sect B 67(3(684)):277–280

Eva Thorn (2001) National programmes in a regional contex. Biologija 4:4–6

Tyler BF, Choriton KH, Thomas ID (1984) Characterization of collected L. *perenne* populations. In: Report, Welsh Plant Breeding Station, Aberystwyth, UK, pp 29–32

Tyler B (1987) Collection, characterization and utilization of genetic resources of temperate forage grass and clover. IBPGRI, pp 1–65

Tucak M, Cupıc T, Popovıc S, Stjepanovıc M, Gantner R, Meglıc V (2009) Agronomic evaluation and utilization of red clover (*Trifolium pratense* L.) germplasm. Not Bot Hort Agrobot Cluj 37(2):206–210

Vilčinskas E (2010) Assessment of clover species (*Trifolium* spp.) diversity in Lithuania: agrobiological and genetic aspect. Summary of doctoral dissertation, Akademija, 26 p

Weibull J (2001) The Nordic-Baltic plant genetic resources project 1998–2000. Biologija 4:12–14

Yonezawa K (1985) A definition of the optimal allocation of effort in conservation of plant genetic resources-with application to sample size determination for field collection. Euphytica 34:345–354

Breeding for Improved Seed Yield
of Red Clover

G. Petrauskas[(⊠)], J. Mikaliūnienė, E. Norkevičienė, G. Statkevičiūtė,
and V. Kemešytė

Institute of Agriculture, Lithuanian Research Centre for Agriculture and Forestry,
Instituto a. 1, Kėdainiai District, Akademija, Lithuania
giedrius.petrauskas@lammc.lt

Abstract. Red clover (*Trifolium pratense* L.) is valued for crude protein and
water soluble carbohydrates content, dry matter digestibility and other properties. On the contrary poor seed yield is often disappointing. Seed yield is very
important criterion in red clover breeding programs, since it is crucial for the
commercialisation of cultivars on the market. Therefore this study was aimed to
compare the seed yield of different red clover cultivars and the most promising
breeding lines. The findings revealed that seed yield decreased more than half in
the second year of harvest. Only a few diploid Lithuania-bred cultivars tended to
produce seed yield significantly higher than average (0.35 t ha^{-1}). The seed
yield of breeding line No. 2114 was found to be lower than that of the other
breeding lines; however, it was the only one stable in both years of harvest.
Therefore breeding line No. 2114 is valuable as breeding material and can be
useful in subsequent stages of selection.

Keywords: *Trifolium pratense* · Seed yield · Breeding line

1 Introduction

Red clover (*Trifolium pratense* L.) is a valuable short-term forage crop of *Fabaceae* L.
family. It is characterized by rapid spring growth and low winter hardiness, which
contributes to its short life span. Like other forage legumes red clover is valued due to
its relatively high crude protein levels, water soluble carbohydrates, dry matter
digestibility and biologically active substances (Boller et al. 2010). Seed yield is relevant both for breeding programs and the market (Amdahl et al. 2017). It has been
reported that seed yield of red clover depends on its ploidy level, which could be either
diploid (2n = 2x = 14) or tetraploid (2n = 4x = 28) (Büyükkartal 2008; Liatukas and
Bukauskaitė 2012). Over decades of breeding, more persistent, higher yielding and
more disease resistant cultivars of this legume have been bred (Svirskis and Juknevičius 2006). Unfortunately, breeding progress for higher seed yield is slow and seed
production in different cultivars is often unsatisfactory (Vleugels et al. 2014, 2016).
Thus there is a need for more information on breeding practices which would ensure
higher seed yield in this forage legume. The purpose of this study was to compare the
seed yield of red clover cultivars and breeding lines.

© Springer International Publishing AG, part of Springer Nature 2018
G. Brazauskas et al. (Eds.): *Breeding Grasses and Protein Crops in the Era of Genomics*, pp. 96–100, 2018.
https://doi.org/10.1007/978-3-319-89578-9_17

2   Materials and Methods

2.1   Experimental Material and Design

Research was carried out at Institute of Agriculture, Lithuanian Research Centre for Agriculture and Forestry (LAMMC ŽI) in period 2012–2014. The study involved 33 cultivars of red clover (*Trifolium pratense* L.) bred in different countries and 30 breeding lines of Lithuanian origin. *Endocalcari – Epihypogleyic Cambisol* (RDg4-k2) with a texture of moderately heavy loam predominates in the fields. Each genotype was sown in plot trial with three replications in randomized complete block design using a manual seeder "Wintersteiger" in 2012. The size of the basic plot was 5 m^2 with 1 m distance between plots. Seed yield was determined in the first and second years (2013–2014) of trial. Ripened seeds of each plot were threshed, dried, cleaned, and weighed (g).

2.2   The Statistical Evaluation of the Research Data

The data of the seed yield from the field was recalculated into t ha^{-1}. Mean and coefficient of variation (CV %) were used to describe the variability of seed yield. The experimental data were processed by using analysis of variance (SAS Institute Inc.). Least significant difference from the average of the seed yields of all cultivars and breeding lines was computed at 95% (LSD_{05}) and 99% (LSD_{01}) confidence levels.

3   Results and Discussion

Variation of the seed yield was observed among 15 diploid and 18 tetraploid cultivars and 15/15 (diploid/tetraploid) breeding lines of two years of harvest. The average seed yield of diploid cultivars amounted to 0.35 t ha^{-1} in 2013, while it was twice as low (0.16 t ha^{-1}) in 2014 (Table 1). This markedly lower seed yield in the second year (2014) resulted from the poor overwinter survival and weather conditions which favoured the occurrence of root and leaf diseases (Mikaliūnienė et al. 2015). The highest seed yield was established for the cultivars of Lithuanian origin during the first year of harvest. This suggests that the origin of cultivars is important for seed yield, because original breeding material from wild populations is naturally adapted to biotic and abiotic stresses. Nevertheless, in the second year of harvest, none of the cultivars showed significantly higher (LSD_{05}) seed yield than the average 0.16 t ha^{-1}.

Breeders and researchers have documented that diploids produce higher seed yields than tetraploids (Boller et al. 2010). There are some studies on different components and factors that may cause a lower seed yield in tetraploid cultivars (Amdahl et al. 2017). To our knowledge, there are not any published data on the seed yield of wild populations of red clover *in situ*. It is believed that investigation of wild populations can lead to a discovery of a genotype exhibiting a stable yield in both years of harvest. In our study, tetraploid cultivars outyielded diploid ones (Table 2). The average seed yield of tetraploid cultivars was by about 26% lower than that of diploid cultivars in the first year and by 19% in the second harvest year. Boller et al. (2010) have reported that tetraploid cultivars produce on average 20–50% lower seed yield than diploid.

Table 1. Seed yield of diploid cultivars of two years of harvest, t ha^{-1}.

Diploid cultivars	Country of origin	1st (2013) year		2nd (2014) year	
		Mean	CV %	Mean	CV %
Arimaičiai	Lithuania (LT)	0.46*	15.23	0.19	50.54
Astra	Bulgaria (BG)	0.36	6.74	0.14	48.90
Britta	Sweden (SE)	0.21*	80.61	0.12	15.16
Jancis	Latvia (LV)	0.36	8.28	0.20	22.56
Jogeva 205	Estonia (EE)	0.27	19.66	0.13	28.11
Jogeva 433	Estonia (EE)	0.24*	46.72	0.15	66.22
Kamaniai	Lithuania (LT)	0.45*	3.44	0.16	57.60
Liepsna	Lithuania (LT)	0.28	16.14	0.15	16.16
Marita	Latvia (LV)	0.38	9.86	0.18	26.91
Palma	Sweden (SE)	0.35	11.53	0.12	47.42
Radviliai	Lithuania (LT)	0.36	30.17	0.16	22.16
Sabtoron	Bulgaria (BG)	0.35	21.10	0.10	26.96
Sandis	Latvia (LV)	0.38	11.62	0.16	38.37
Spurt	Czech Republic (CZ)	0.34	12.26	0.20	48.52
Vyčiai	Lithuania (LT)	0.40	10.94	0.18	45.28
Average		**0.35**		**0.16**	
*LSD$_{05}$		0.051		0.064	

Table 2. Seed yield of tetraploid cultivars of two years of harvest, t ha^{-1}.

Tetraploid cultivars	Country of origin	1st (2013) year		2nd (2014) year	
		Mean	CV %	Mean	CV %
Blizard	Czech Republic (CZ)	0.31*	5.91	0.20*	43.15
Cyklon	Czech Republic (CZ)	0.21	11.72	0.14	23.06
Dolly	Czech Republic (CZ)	0.22	22.26	0.08	58.19
Jlte	Estonia (EE)	0.34**	8.76	0.11	29.07
Kaive	Latvia (LV)	0.20*	19.82	0.15	3.46
Kiršinai	Lithuania (LT)	0.25	21.46	0.15	15.11
Ostro	Estonia (EE)	0.27	11.31	0.20*	8.46
Sadūnai	Lithuania (LT)	0.34**	8.51	0.12	47.27
Sara	Sweden (SE)	0.13*	31.80	0.09	147.25
Skriveru tetra	Latvia (LV)	0.23	18.33	0.11	18.57
Sprint	Czech Republic (CZ)	0.26	24.59	0.15	18.59
Start	Czech Republic (CZ)	0.40**	4.86	0.22**	41.82
Temasa	Switzerland (CH)	0.20*	53.06	0.04*	80.79
Tempus	Czech Republic (CZ)	0.25	7.77	0.17	7.13
Triton	Sweden (SE)	0.23	15.77	0.04*	52.45
Varte	Estonia (EE)	0.34**	8.76	0.11	107.90
Vyliai	Lithuania (LT)	0.12*	34.50	0.07	55.85
Vulkan	Czech Republic (CZ)	0.32*	6.33	0.18	15.04
Average		**0.26**		**0.13**	
*LSD$_{05}$		0.051		0.066	
**LSD$_{01}$		0.069		0.088	

Amdahl et al. (2017) have distinguished a few components as a possible reason for low seed yield in tetraploids, including number of flower heads, number of florets per flower head, length of the corolla tube, seed number per flower head, fertility.

Seed yield of cultivars was compared with that of breeding lines (Table 3).

Table 3. Seed yield of breeding lines of two years of harvest, t ha^{-1}.

Breeding line (2n)	1st (2013) year,		2nd (2014) year		Breeding line (4n)	1st (2013) year		2nd (2014) year	
	Mean	CV %	Mean	CV %		Mean	CV %	Mean	CV %
2086	0.17**	38.18	0.13	69.13	2293	0.22	5.81	0.06	98.83
2087	0.23	30.76	0.13	58.50	2300	0.21	5.74	0.03	26.91
2088	0.21	19.61	0.04	16.43	2286	0.05**	53.13	0.03	91.50
2091	0.22	30.84	0.07	64.97	2301	0.27*	17.35	0.03	77.57
2093	0.24	16.01	0.10	43.44	2268	0.21	12.84	0.08	123.35
2094	0.26	23.79	0.05	67.83	2298	0.32**	8.69	0.04	40.70
2102	0.27	24.14	0.14	60.97	2190	0.08**	48.24	0.01	131.49
2105	0.27	12.54	0.06	77.42	2282	0.28**	42.53	0.07	154.23
2106	0.08**	45.22	0.01*	28.72	2302	0.25	8.61	0.06	19.72
2107	0.21	39.45	0.04	72.50	2114	0.14*	58.58	0.10*	72.95
2108	0.25	22.82	0.09	23.46	2296	0.12**	40.65	0.03	90.57
2188	0.32	26.22	0.11	76.78	2274	0.24	39.14	0.01	35.37
2295	0.34**	26.01	0.16	59.96	2096	0.25	8.61	0.01	92.17
2297	0.36**	5.58	0.10	80.39	2098	0.13*	14.61	0.04	137.06
2299	0.37**	8.71	0.14	44.54	2095	0.20	26.97	0.02	82.99
Average	**0.25**		**0.09**			**0.20**		**0.04**	
*LSD$_{05}$	0.067		0.072			0.054		0.05	
**LSD$_{01}$	0.09		0.098			0.073		0.068	

Average seed yield of diploid and tetraploid breeding lines were quite similar in the first year of harvest 0.25 t ha^{-1} (2n) and 0.20 t ha^{-1} (4n), respectively. In the second harvest year seed yield of tetraploid breeding lines were lower more than 50% compared to diploids. Three breeding lines of diploids (2299, 2297, 2295) were found to produce a significantly higher (LSD$_{01}$) seed yield than the average in 2013; however, no statistically significant difference (LSD$_{01}$) was established in the second year of harvest. Our study showed that the three breeding lines of tetraploids (2301, 2298, 2282) produced higher yields than the average (0.20 t ha^{-1}) in 2013, but none of them was so productive in the second year of harvest. The reasons for the low seed yield in the second year of harvest were explained in the previous study (Mikaliūnienė et al. 2015). One of the tetraploid breeding line (2114) exhibited stable seed yields in both years of harvest despite the low seed yield. However, further studies are needed to validate this finding.

4   Conclusion

Seed yield of both diploid and tetraploid red clover was markedly lower in the second year of harvest. Because of the higher winterhardiness, root and leaf disease resistance, the cultivars of local origin were more productive than foreign ones (Mikaliūnienė et al. 2015). However, one of the breeding lines (2114) showed a stable seed yield in both years of harvest. This fact is important for the development of marketing and breeding programs.

Acknowledgments. This study was funded through the long-term research program "Genetics and purposeful change of genotypes of agricultural and forest plants" implemented by Lithuanian Research Centre for Agriculture and Forestry.

References

Amdahl H, Aamlid ST, Marum P, Ergon Å, Alsheikh M, Rognli AO (2017) Seed yield components in single plants of diverse Scandinavian tetraploid red clover populations (*Trifolium pratense* L.). Crop Sci 57:108–117

Boller B, Schubiger FX, Kölliker R (2010) Red clover. In: Boller B (ed) Handbook of plant breeding. Springer, Dordrecht, pp 439–455

Büyükkartal HNB (2008) Causes of low seed set in the natural tetraploid *Trifolium pratense* L. (*Fabaceae*). Afr J Biotech 7:1240–1249

Liatukas Ž, Bukauskaitė J (2012) Differences in yielding capability of diploid and tetraploid red clover in Lithuania. Proc Latvian Acad Sci Sect B 66:163–167

Mikaliūnienė J, Lemežienė N, Danytė V, Supronienė S (2015) Evaluation of red clover (*Trifolium pratense* L.) resistance to Sclerotinia crown and root rot (*Sclerotinia trifoliorum*) in the laboratory and field conditions. Zemdirb Agric 102:167–176

Svirskis A, Juknevičius S (2006) Breeding of red clover varieties for conventional and organic agriculture. Zemdirb Agric 93:314–321

Vleugels T, Ceuppens B, Cnops G, Lootens P, van Parijs DRF, Smagghe G, Roldán-Ruiz I (2016) Models with only two predictor variables can accurately predict seed yield in diploid and tetraploid red clover. Euphytica 209:507–523

Vleugels T, Cnops G, Roldán-Ruiz I (2014) Improving seed yield in red clover through marker-assisted parentage analysis. Euphytica 200:305–320

Genetic Improvement of Quality and Agronomic Traits

Breeding Forage and Grain Legumes to Increase EU's and China's Protein Self-sufficiency

B. Julier[1]([⌧]), L. Skøt[2], S. Weise[3], Đ. Karagić[4], I. Roldán-Ruiz[5], P. Barre[1], and D. Lloyd[2]

[1] P3F, INRA, 86600 Lusignan, France
bernadette.julier@inra.fr
[2] Institute of Biological, Environmental and Rural Sciences,
Aberystwyth University, Aberystwyth, UK
[3] Leibniz Institute of Plant Genetics and Crop Plant Research (IPK),
Gatersleben, 06466 Stadt Seeland, Germany
[4] Forage Crops Department, Institute of Field and Vegetable Crops IFVCNS,
Novi Sad, Serbia
[5] Plant Sciences Unit, Institute of Agriculture and Fisheries Research ILVO,
Melle, Belgium

Abstract. Both Europe and China have a low level of protein autonomy. They import 30% and 60%, respectively, of the world plant protein production for both human and animal nutrition. However, in both continents, regions with appropriate agro-ecological conditions for the production of plant proteins exist. The project EUCLEG, recently granted by the H2020 programme of European Union, has the ambition to give a new momentum to forage legume (alfalfa and red clover) and grain legume (pea, faba bean and soybean) breeding. The scientific objectives of this project are (i) to broaden the genetic base of legume crops, (ii) to analyse the genetic diversity of European and Chinese legume accessions using phenotypic traits (yield, quality in a context of climate change) and molecular markers, (iii) to analyse the genetic architecture of key breeding traits using association genetics (GWAS) and (iv) to evaluate the benefits brought by genomic selection (GS) to create new legume varieties. Comprehensive searchable databases containing passport data, as well as agronomic and genetic features will be developed. The breeders that are members of EUCLEG will exploit the results to create the new varieties needed to improve protein production.

Keywords: Forage legume · Grain legume · Association genetics
Genomic selection · Plant genetic resources · Biodiversity informatics

1 Introduction

Protein dependency in Europe and China originates from insufficient production of protein rich crops in these two continents. For the past three to four decades, Europe has imported 65% of its protein (approximately 35 million tonnes of soybean per year).

© Springer International Publishing AG, part of Springer Nature 2018
G. Brazauskas et al. (Eds.): *Breeding Grasses and Protein Crops in the Era of Genomics*, pp. 103–108, 2018.
https://doi.org/10.1007/978-3-319-89578-9_18

China imported 74 million tonnes of soybean in 2014 (corresponding to 60% of world market trade). Since the year 2000, the Chinese protein demand has increased steadily at a rate of 6 million tonnes per year because of increased demand for animal products. A higher level of independency from imports can be achieved through increased local production of plant protein. Legume crops are ideal for protein production as this is mostly based on symbiotic fixation of atmospheric nitrogen. This has a positive environmental impact as well as providing a number of other ecosystem services (energy saving, improvement of soil structure and composition, reduction of greenhouse gas emission, hosted biodiversity). The project EUCLEG, recently granted by the H2020 programme of European Union, will target five major legume crops (alfalfa, red clover, faba bean, pea and soybean) selected for their agronomic properties, economic relevance and current use in Europe and China. Grain legumes have a major role in EUCLEG, however, the project will focus particularly on forage crops that can deliver up to 2.3 tonnes protein per hectare per year while grain legumes reach up to 1.3 tonnes/ha. Indeed, the development of legume-based grassland–livestock systems undoubtedly constitutes one of the pillars for more sustainable and competitive ruminant production systems.

2   Concept

The EUCLEG concept combines: (i) development of genomic tools in forage species and faba bean needed to analyse the genetic architecture of agronomic traits; (ii) evaluation of genetic resources including natural and selected populations useful for adaptation of species to current or new constraints or uses; (iii) dense genotyping of all populations that will be phenotyped; (iv) establishment of searchable databases gathering passport, phenotypic and genotypic data of genetic resources; (v) identification of genes and markers involved in phenotypic variation based on phenotypic traits and genomic data; (vi) estimation of genomic selection efficiency.

3   Approach

3.1   Plant Material and Phenotyping

Using diverse genetic resources and taking advantage of advanced molecular tools, EUCLEG aims at identifying and developing the best genetic resources, phenotyping methods and molecular tools to breed legume varieties with improved performance under biotic and abiotic stresses in the representative European and Chinese agro-ecological areas.

EUCLEG is structured around five research and innovation work packages (WP) and additionally one dissemination work package and one management work package (Fig. 1).

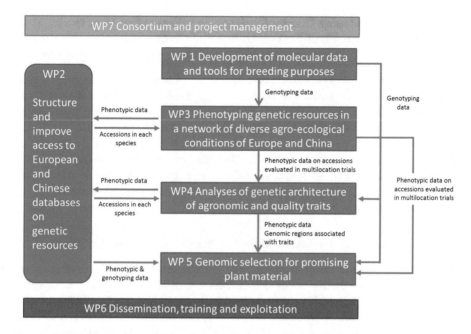

Fig. 1. Distribution of work in the seven work packages (WP) of EUCLEG and interdependency

3.2   WP1 – Development of Molecular Data and Tools for Legume Crop Breeding

WP1 will produce genome sequences and identify SNP polymorphisms for forage legumes and faba bean, using high throughput genotyping methods essential for genomics assisted breeding. Such data and tools are already available for pea and soybean. These newly developed and existing genomic resources and tools will be used in WP1 to generate genotypic information for all the germplasm considered in EUCLEG. This information will be exploited in WP3 to analyse the diversity and population structure in all five species, in WP4 to identify marker-trait associations and in WP5 for genomic selection.

3.3   WP2 – Structure and Improve Access to European and Chinese Databases on Genetic Resources

WP2 will consolidate existing databases and build new ones describing the European and Chinese collections of the five legumes, and improve access to genetic resources. The phenotypic and genotypic data generated in WP1, WP3, WP4 and WP5 will be included in these databases. Gaps in existing plant genetic resources databases will be identified. WP2 will also develop agreements for data sharing and genetic resources exchanges between project partners (Europe and China) consistent with their international legal commitments and national laws.

3.4   WP3 – Phenotyping Genetic Resources in a Network of Diverse Agro-Ecological Conditions of Europe and China

WP3 will provide the user community high quality evaluation data on a wide range of plant genetic resources. Trials connected by the use of a set of common control varieties across a network of well-characterised agro-ecological areas will provide an extensive description of European and Chinese germplasm and an overview of genotype x environment interaction. In addition, WP3 will describe the genetic structure of this germplasm using molecular markers. Data will be used to improve the description of genetic resources in databases (WP2). The data collected on accessions studied in WP3 will also be part of WP4 and WP5 data for analysis of genetic determinism of agronomic traits, calculation of genomic selection equations and identification of populations of agronomic interest for development of promising material.

3.5   WP4 – Analyses of Genetic Control of Agronomic and Quality Traits

WP4 will gain a better understanding of the genetic architecture of important agronomic traits using association mapping approaches, which in turn will deliver improved breeding tools and approaches. WP4 will identify genes and markers that explain a large part of the phenotypic variation in traits of agronomic and biological significance, and compare results across species. WP4 will also provide input to WP5.

3.6   WP5 – Genomic Selection for Promising Plant Material

WP5 will provide knowledge and tools for breeders to implement genomic selection in their breeding programmes. User-friendly software will be developed to derive genomic breeding values. By using the data collected in WP3 and WP4 as well as phenotypic and genotypic information obtained by breeders on their own material, WP5 will gather a sufficient amount of data to establish models for predicting breeding values. WP5 will also identify promising genotypes for breeding.

3.7   WP6 – Outreach

WP6 will disseminate the project outcomes to the scientific community and transfer the innovations to stakeholders. It aims to ensure the effective and efficient distribution of EUCLEG results and findings to the target audience including, but not restricted to, theoreticians and practitioners in plant breeding, legume growers, legume seed traders, legume consumers, NGOs, and policymakers.

3.8   WP7 – Consortium Coordination and Project Management

WP7 will manage the administrative, logistical, and financial coordination, including the monitoring of progress and resources consumption against milestones.

4 Partnership

Thirty-eight (26 European and 12 Chinese) partners are members of the EUCLEG consortium (Fig. 2, Table 1). Seventeen (17) European and 11 Chinese public institutes or universities share interest in legume genetics, genomics and genetic resources management. Nine European and one Chinese breeding companies will use knowledge

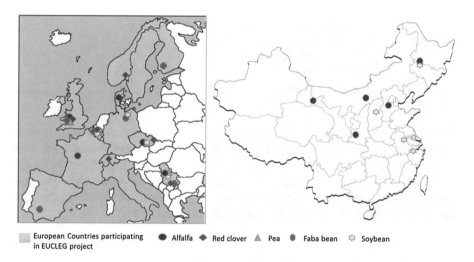

European Countries participating in EUCLEG project   ● Alfalfa   ◆ Red clover   ▲ Pea   ● Faba bean   ✿ Soybean

Fig. 2. Partnership in EUCLEG. Points indicate the locations of field trials

Table 1. Partners of EUCLEG

	Public	Private
Europe	INRA (FRA)	Agro Seed Research (BEL)
	Agroscope (CHE)	Agrovegetal (ESP)
	ART (CZE)	Barenbrug (FRA)
	IFVCNS (SRB)	Boreal (FIN)
	IKBKS (SRB)	DLF Seeds (DK)
	AU-IBERS (GBR)	Graminor (NOR)
	IFAPA (ESP)	Jouffray-Drillaud (FRA)
	IPGRI (ITA)	Lantmännen (SWE)
	JKI (DEU)	RAGT (FRA)
	IPK (DEU)	Progeno (BEL)
	NordGen (SWE)	
	NMBU (NOR)	
	ILVO (BEL)	
	INRA-Transfert (FRA)	
	IPGRI (ITA)	
	U-Ghent (BEL)	
	VIB (BEL)	

(*continued*)

Table 1. (*continued*)

	Public	Private
China	CAAS-IGR (Inner Mongolia) HAAS (Heilongjiang) IMAU (Inner Mongolia) CAAS-IAS (Beijing) JAAS (Jiangsu) NJAU (Nanjing) NEAU (Harbin) CAS-IGA (Changchun) NWAFU (Yangling) SXAU (Shanxi) ZAAS (Zhejiang)	Jiuquan Daye Seed Industry

gained on phenotypic variation among populations, molecular tools, genetic diversity and breeding methodologies to increase the genetic progress in their breeding schemes. In the long term, improved varieties adapted to most agro-ecological conditions and uses will be registered. A European SME will develop a software to handle passport, phenotypic and genetic data collected on legume populations, from data deposition and search to genome-wide association tools and genomic selection.

Acknowledgments. EUCLEG is a RIA project (n°727312) granted by European Union, call H2020-SFS-2016-2017, topic SFS-44-2016. Chinese groups are supported by the Chinese Ministry of Agriculture (MOST).

Identification of Loci Controlling Timing of Stem Elongation in Red Clover Using GBS of Pooled Phenotypic Extremes

Å. Ergon[1]([⊠]), Ø. W. Milvang[1], O. Milvang[2], and M. R. Kovi[1]

[1] Faculty of Biosciences, Norwegian University of Life Sciences, P.O.
Box 5003, 1432 Ås, Norway
ashild.ergon@nmbu.no
[2] Brennaveien 143, 1279 Oslo, Norway

Abstract. A population of red clover (*Trifolium pratense* L.), was characterized for timing of stem elongation. Of the 672 plants tested, 52 of the earliest
and 52 of the latest were chosen for genetic characterization using genotyping
by sequencing (GBS). The 52 individuals from each group were randomly
divided into three subgroups and equal amounts of DNA from each individual in
each subgroup was combined in a pool, creating a total of six pools, that were
subject to GBS. Single nucleotide polymorphisms (SNPs) with allele frequencies that differed significantly between the early and the late group were identified by the use of pairwise F_{ST}-values for each SNP. At the highest stringency,
we found 57 biallelic SNPs with significantly different allele frequencies in the
early and the late groups, of which 37 could be mapped to a chromosome and
the rest to unplaced scaffolds ($P < 0.01$, false discovery rate < 0.001). The 37
SNPs corresponded to 22 loci when defining a locus as a ±500 Kb region
surrounding a SNP, an interval known to be associated with an almost complete
decay of linkage disequilibrium in the studied population. Significant SNPs
were distributed among all chromosomes, with the highest number on chromosome 6 and 7.

Keywords: *Trifolium pratense* · Earliness · SNP

1 Introduction

Stem elongation is important for radiation capture, competitive ability and herbage and
seed yields in red clover, but reduces its digestibility and energy concentration as
forage. A better understanding of the genetic control of initiation of stem elongation
can aid our breeding efforts to optimize timing of stem elongation. The aim of this
study was to (1) identify SNP markers and chromosomal regions controlling timing of
stem elongation in a Norwegian red clover variety and (2) investigate whether it is
possible to do this by a "genotyping of pooled phenotypic extremes"-approach. We did
this by screening a large number of individuals, creating replicate groups of extremely
early or extremely late individuals, genotyping of the corresponding DNA pools, and
identification and mapping of SNPs with significantly different variant allele frequencies in early vs. late groups.

© Springer International Publishing AG, part of Springer Nature 2018
G. Brazauskas et al. (Eds.): *Breeding Grasses and Protein Crops in the Era of Genomics*, pp. 109–112, 2018.
https://doi.org/10.1007/978-3-319-89578-9_19

2 Material and Methods

Seeds of red clover (*Trifolium pratense* L., cultivar 'Lea' (Graminor Ltd.)) were sown in a greenhouse in September 2015 and grown at approximately 16 °C with a 20 h photoperiod (natural light supplied with 90 μmol m^{-2} s^{-1} PAR (HPQ/HTI-P lamps)). The number of days from sowing until the first elongating stem was 2 cm long (days to elongation – DTS) was recorded for each plant. Of the 672 plants tested, 52 of the earliest (average DTS of 37.7) and 52 of the latest (average DTS of 80.5) were chosen for genetic characterization. The 146 plants that did not elongate during the course of the experiment were excluded.

DNA was extracted from leaf tissue of each of the 104 selected individuals. The 52 individuals in each of the two phenotypic groups were randomly divided into three subgroups and equal amounts of DNA from each individual in each subgroup was combined in a pool, creating a total of six pools, each of which were distributed into 15–16 wells on a 96-well plate. GBS library preparation and sequencing, as well as SNP calling, was done by the Cornell University Biotechnology Resource Center according to Elshire et al. (2011). Both Pst1- and ApeK1-digested libraries were sequenced. The 'discovery' pipeline (Glaubitz et al. 2013) in TASSEL 3.0 was used to call SNPs from the sequenced GBS libraries and locate them on the red clover genome (De Vega et al. 2015).

After removing SNP and well combinations were the maximum read number (127) had been reached for one or both variants, allele frequencies were calculated for each SNP and pool by summing up the number of reads in all wells. After removing SNPs with missing data > 0.1 or a minor allele frequency < 0.05 across all wells, as well as SNPs with less than a total of 100 reads for one or more of the six pools, 3590 (Pst1) and 43242 (ApeK1) SNPs remained for analysis. This equals an average distance between SNPs of 117 and 9.7 kb, respectively. Seven percent of the SNPs were tri-allelic. These were excluded from the further analysis reported here.

For each SNP, pairwise F_{ST} values $(\frac{\overline{q^2}-\bar{q}^2}{\bar{q}(1-\bar{q})})$ were calculated for all pools against the average allele frequency of the three pools of the contrasting phenotype (a total of six F_{ST} values). A chi-square test was used to identify significant F_{ST}'s at $P < 0.1$, $P < 0.05$ and $P < 0.01$ levels, using the test statistic $x^2 = 2NF_{ST}$, where $2N$ = the sum of genotyped gametes in the two populations (Hedrick 2011). Only SNPs with significant F_{ST}'s for all six comparisons were regarded as having significantly different allele frequencies in the early vs. the late phenotypic group. Corresponding estimates of the false discovery rate (*FDR*) were calculated for each *P*-level as $\frac{l*P^3}{d}$, where l = number of SNP loci tested and d = the number of SNP loci identified as significant.

3 Results and Discussion

A significance level of P < 0.05 gave a satisfactory FDR of 0.01 (Pst1) and 0.03 (ApeK1), and identified 37 (Pst1) and 160 (ApeK1) SNPs with different allele frequencies in early vs. late groups, of which a total of 128 was mapped to a chromosome (Table 1). The corresponding numbers at P < 0.01 were 15 (Pst1), 42 (ApeK1) and 37 (total, mapped to chromosomal location). There is very little genetic structure and linkage disequilibrium in the studied population. In a previous study, the two first principal components in a PCA of SNP data for 86 individuals explained only a few percent of the genetic variation (De Vega et al. 2015). In the same study, the average linkage disequilibrium (r^2) at 100 Kb, calculated for each chromosome separately, ranged between 0.19 and 0.25. At 500 Kb the linkage disequilibrium had almost decayed completely (r^2 around 0.05). When regarding SNPs within ±100 Kb, or within ±500 Kb, as one locus, the SNPs with an effect on earliness (P < 0.05) grouped into 85 and 61 loci, respectively (Table 2). At a significance level of P < 0.01, the equivalent numbers were 26 and 22. Chromosome 6 was the chromosome with the highest number of loci with an effect on earliness, followed by chromosome 7. Chromosome 6 is largely syntenic with *Medicago truncatula* chromosome 7 (De Vega et al. 2015), on which a very consistent QTL for flowering time as well as several

Table 1. The number of SNPs with significantly different allele frequencies in early vs. late elongating phenotypic groups at three significance levels.

Enzyme	Number of biallelic SNPs tested	Number of SNPs with significantly different allele frequencies in early vs. late groups (FDR)		
		P < 0.1	P < 0.05	P < 0.01
Pst1	3356	55 (0.06)	37 (0.01)	15 (0.0002)
ApeK1	40387	321 (0.1)	160 (0.03)	42 (0.001)

FDR - false discovery rate

Table 2. The chromosomal distribution of SNPs with significantly different allele frequencies in early vs. late elongating phenotypic groups at P < 0.05 and FDR = 0.01 (Pst1) or 0.03 (ApeK1). Corresponding numbers for P < 0.01 are given in parentheses. SNPs that were located within either ±100 Kb or ±500 Kb were considered as one locus.

Chromosome	Number of SNPs with significantly different allele frequencies in early vs. late groups			Number of loci	
	Pst1	ApeK1	Total	±100 Kb	±500 Kb
1	3 (1)	11 (0)	14 (1)	10 (1)	10 (1)
2	2 (1)	6 (0)	8 (1)	6 (1)	6 (1)
3	0 (0)	8 (3)	8 (3)	5 (2)	5 (2)
4	5 (0)	8 (1)	13 (1)	10 (1)	7 (1)
5	1 (0)	14 (3)	15 (3)	9 (2)	7 (1)
6	6 (2)	36 (15)	42 (17)	27 (11)	14 (9)
7	6 (6)	23 (6)	28 (11)[a]	18 (8)	12 (7)
Unplaced scaffolds	14 (5)	54 (14)	68 (19)	-	-

[a]One SNP was detected by both enzymes

flowering time-related genes are located (Pierre et al. 2008). One of the loci we identified in our study is located 32 Kb away from a *CONSTANS*-like gene, which could be syntenic to the one identified as a candidate for the QTL in *M. truncatula* (Pierre et al. 2011).

4 Conclusion

Genotyping of pooled phenotypic extremes within a population allowed us to identify a number of SNP markers and QTLs for earliness of stem elongation in red clover.

Acknowledgments. This study was funded by the Norwegian Research Council (project AGROPRO – grant agreement number 225330).

References

De Vega JJ, Ayling S, Hegarty M, Kudrna D, Goicoechea JL, Ergon Å, Rognli OA, Jones C, Swain M, Geurts R, Lang C, Mayer KFX, Rössner S, Yates S, Webb KJ, Donnison IS, Oldroyd GED, Wing RA, Caccamo M, Powell W, Abberton MT, Skøt L (2015) Red clover (*Trifolium pratense* L.) draft genome provides a platform for trait improvement. Sci Rep 5 (Article no. 17394)

Elshire RJ, Glaubitz JC, Sun Q, Poland JA, Kawamoto K, Buckler ES, Mitchell SE (2011) A robust simple genotyping-by-sequencing (GBS) approach for high diversity species. PLoS One 6(5):e19379

Glaubitz JC, Casstevens TM, Lu F, Harriman J, Elshire RJ, Sun Q, Buckler ES (2013) TASSEL-GBS: a high capacity genotyping by sequencing analysis pipeline. PLoS One 9(2): e90346

Hedrick PW (2011) Genetics of populations, 4th edn. Jones and Bartlett Publishers, Sudbury, 675 p. ISBN-13: 978-0-7637-5737-3

Pierre J-B, Huguet T, Barre P, Huyghe C, Julier B (2008) Detection of QTLs for flowering date in three mapping populations of the model legume species *Medicago truncatula*. Theor Appl Genet 117:609–620

Pierre J-B, Bogard M, Herrmann D, Huyghe C, Julier B (2011) A *CONSTANS*-like gene candidate that could explain most of the genetic variation for flowering date in *Medicago truncatula*. Mol Breed 28:25–35

Grain Yield and Morphology of Dwarf *vs* Tall White Lupin in Mediterranean Environments

P. Annicchiarico[1], A. M. Carroni[2], P. Manunza[2], C. Huyghe[3], and L. Pecetti[1(✉)]

[1] Research Centre for Fodder Crops and Dairy Productions (CREA-FLC), Lodi, Italy
luciano.pecetti@crea.gov.it
[2] Research Unit for Mediterranean Agro-Pastoral Systems (CREA-AAM), Sanluri, Italy
[3] Institut National de Recherche Agronomique (INRA), Lusignan, France

Abstract. There is lack of information on the preferable plant architecture of white lupin (*Lupinus albus* L.) for Mediterranean environments. The objective of this study was to compare tall *vs* dwarf near-isogenic lines from two gene pools for grain yield and plant morphology. The four lines were assessed in a Mediterranean-climate site of Sardinia for two seasons under two sowing densities (25.4 and 35.5 seeds/m^2). On average, one gene pool (#15) exhibited a semi-determinate growth habit on the basis of its high seed proportion on the main stem relative to the other, indeterminate gene pool #16 (54.9% *vs* 45.5; $P < 0.05$). As expected, dwarfism reduced plant stature (106 *vs* 124 cm) and increased the harvest index (0.293 *vs* 0.250). Compared with the tall variant, dwarfism led to significantly ($P < 0.05$) higher grain yield in the gene pool #16, and lower yield in the gene pool #15, giving rise to significant plant stature × gene pool interaction ($P < 0.05$). Sowing densities and seasons did not differ for grain yield, nor interacted significantly with plant stature or gene pool for this variable ($P > 0.05$). Although preliminary, our results for Mediterranean environments highlight the genotype-specific usefulness of dwarfism and suggest that this trait might be beneficial in germplasm with indeterminate growth habit.

Keywords: Harvest index · *Lupinus albus* · Ideotype · Near-isogenic line
Plant architecture

1   Introduction

White lupin has increasing interest in Europe to reduce the marked deficit of high-protein feedstuff. In soils to which it is adapted (neutral to acid pH; active lime < 1%), it is the cool-season grain legume with highest protein yield per unit area in rain-fed environments of southern Europe (Annicchiarico 2008). There is renewed interest in white lupin also for food use, owing to outstanding seed protein content and oil quality (Arnoldi 2005; Boschin et al. 2007).

White lupin breeders can rely on agronomically-outstanding, largely untapped landrace genetic resources (Annicchiarico et al. 2010), but need to define the best plant architecture for their target region. White lupin usually shows an indeterminate growth

© Springer International Publishing AG, part of Springer Nature 2018
G. Brazauskas et al. (Eds.): *Breeding Grasses and Protein Crops in the Era of Genomics*, pp. 113–117, 2018.
https://doi.org/10.1007/978-3-319-89578-9_20

habit, with continued flowering and branching from main stem to primary, secondary or higher-order branches (Noffsinger et al. 2000). Yield instability and risk of lodging are major drawbacks associated with indeterminate habit (Julier et al. 1993; Annicchiarico et al. 2010). Determinate variants with restricted vegetative development have been exploited by breeders (Julier et al. 1993), although preliminary results across climatically contrasting environments suggested limited potential of determinate white lupin in Italy (Annicchiarico and Carroni 2009). Dwarfism is another genetic feature preventing excessive vegetative development in indeterminate white lupin, by reducing the length of main stem and branches (Harzic and Huyghe 1996). Comparisons of dwarf *vs* tall germplasms were carried out in temperate environments of western Europe (Harzic and Huyghe 1996; Harzic et al. 1995, 1996), while information is lacking for Mediterranean environments.

The objective of this study was comparing dwarf *vs* tall near-isogenic lines for grain yield and plant morphology under Mediterranean conditions, assessing the consistency of results across different gene pools, plant densities and cropping seasons.

2 Material and Methods

Tall and dwarf near-isogenic lines of two gene pools (coded as #15 and #16) developed at INRA were grown in Sanluri (Sardinia, Italy, 39°30′ N, 8°50′ E) for two seasons (2006-07 and 2007-08) in rain-fed autumn-sown field experiments. Within a typical Mediterranean climate, both seasons had more favourable spring rainfall than average (151 mm and 117 mm from April to June, respectively, *vs* a long-term value of 88 mm). Also, the second season was slightly colder than average (10 frost days *vs* 2 in the long term; 13.8 °C daily mean temperature from October to June *vs* 15.0 °C). Soil pH (<7.5) and active lime (<0.4%) were suitable for lupin cropping.

The lines were sown at two sowing densities in plots with five rows, 157 cm long and 37.5 cm apart, according to a factorial randomized complete block design. The higher sowing density was 35.5 seeds/m^2, obtained by sowing 21 seeds spaced 7.5 cm per row; the lower density was 25.5 seeds/m^2, by sowing 15 seeds spaced 10.5 cm per row. Before sowing, seeds were inoculated with a commercial rhizobium (NPPL HiStick, Becker Underwood, Toulouse, France), to ensure proper nodulation and N fixation. Soil analyses suggested no fertilization requirement.

Dry grain yield was recorded on plants from the three central rows of each plot (1.8 m^2), assessing seed moisture and individual seed dry weight on a random sample of 200 seeds. All remaining traits were assessed on a subsample of six plants per plot. They included dry grain yield on the main stem and on branches and aerial biomass following oven drying and, from these data, harvest index and proportion of grain on the main stem. Further traits recorded in both seasons were plant height as the natural height of the main stem inclusive of inflorescence, and internode length as average of three internodes just beneath the terminal inflorescence on main stem. Number of leaves on the main stem was recorded in the second season.

A combined analysis of variance (ANOVA) including the factors stature (tall or dwarf), gene pool (#15 or #16), density, season, and block within season was performed for all traits. A second ANOVA included the factors near-isogenic line (four lines), density, season, and block within season. Analyses for the number of leaves did not include the factor season.

3 Results and Discussion

The first cropping season produced plants with taller stature (132 *vs* 98 cm), longer internodes (33.8 *vs* 29.9 cm), more aerial biomass (7.66 *vs* 5.58 t/ha), lower harvest index (0.224 *vs* 0.319) and lower proportion of seed on the main stem (38.5 *vs* 61.8%) than the second season ($P < 0.05$). The greater plant vegetative development in the first season was probably due to slower vernalization caused by milder winter, as reported also by Harzic et al. (1995), besides somewhat more favourable growing conditions arising from slightly wetter spring. However, seasons did not differ for grain yield, and no significant interaction of season with plant stature, sowing density or gene pool was detected ($P > 0.05$). The two sowing densities did not differ for any trait across seasons (data not shown), and showed no interaction with plant stature or gene pool. Comparing a dwarf *vs* a tall genotype under irrigation, Harzic et al. (1995) found that contrasting plant densities modified the yield distribution on different branch orders but did not affect consistently the grain yield.

On average, the gene pool #15 was characterized by main stems with somewhat shorter internodes (29.9 *vs* 33.7 cm), more leaves (25.4 *vs* 21.4) and higher proportion of seed (54.9% *vs* 45.5) than the gene pool #16 ($P < 0.05$). The extent of seed allocation to the main stem exhibited by the pool #15 would classify this material as semi-determinate (Julier et al. 1993), whereas the pool #16 features an indeterminate growth habit.

The introduction of dwarfism caused similar morphological changes in the two genetic backgrounds. Table 1 reports values of the four near-isogenic lines averaged across seasons and sowing densities. Dwarfism reduced plant height by 15% (106 *vs* 124 cm) and increased the harvest index by 17% (0.293 *vs* 0.250) compared with tall stature ($P < 0.05$). The reduction of plant height was associated with a reduction of about 6% for both internode length and number of internodes (the latter, as inferred from the number of leaves on the main stem). Also, dwarfism tended to increase the proportion of seed on the main stem (+9.8%; $P < 0.10$). Harzic and Huyghe (1996) reported greater reductions of plant stature and internode length than these. The higher proportion of main stem seed in dwarf than in tall lines did not agree with findings by Harzic et al. (1995), who attributed to late pod and seed abortion the lower main stem seed proportion of the dwarf genotype.

The comparison of dwarf *vs* tall material revealed no difference for mean grain yield. Actually, the usefulness of dwarfism depended on the specific gene pool. Compared with the tall variant, dwarfism led to significantly ($P < 0.05$) higher grain yield in the gene pool #16 characterized by an indeterminate growth habit, and lower yield in the gene pool #15 characterized by semi-determinate habit (Table 1), generating significant ($P < 0.05$) plant stature × gene pool interaction. Concurrently, dwarfism provided significantly higher harvest index only in gene pool #16, and lower aerial biomass only in gene pool #15 (Table 1), with a trend towards plant stature × gene pool interaction

116   P. Annicchiarico et al.

Table 1. Traits of white lupin near-isogenic lines differing in plant stature from two gene pools, averaged across two sowing densities and two cropping seasons in a Sardinian location.

Variable[†]	Gene pool #15		Gene pool #16	
	Dwarf	Tall	Dwarf	Tall
Dry grain yield, t/ha	1.55 b	1.86 a	1.93 a	1.54 b
Dry aerial biomass, t/ha	5.42 b	6.99 a	6.43 ab	6.58 ab
Harvest index	0.286 a	0.266 ab	0.300 a	0.234 b
Plant height[‡], cm	106 b	125 a	106 b	123 a
Internode length, cm	29.4 b	30.5 ab	32.4 ab	35.0 a
No. leaves/main stem	24.5 ab	26.3 a	20.9 c	22.0 bc
Proportion of seed on main stem, %	56.9 a	52.8 ab	48.2 ab	42.8 b
Individual seed weight, g	0.265 a	0.272 a	0.286 a	0.270 a

[†] In each row, mean values followed by different letters differ at $P < 0.05$ according to Duncan's multiple range test.
[‡] Natural height of main stem including inflorescence.

also for these variables ($P < 0.10$). In relative terms, dwarfism increased the harvest index by 28% in the gene pool #15 and 7% in #16. Higher harvest index of dwarf material depended more on size reduction of vegetative organs than higher grain yield, in large agreement with what observed by Harzic et al. (1995).

The differences in grain yield between the two pairs of near-isogenic lines tended to be wider in the first, climatically more favourable season, generating significant stature × gene pool × season interaction ($P < 0.05$). For example, dwarfism of the gene pool #16 in the first season led to markedly higher harvest index (+37%) and grain yield (+62%) than its tall variant in this season than in the second one.

4 Conclusion

The contrasting results that we obtained for the two gene pools suggest that dwarfism has germplasm-specific value for Mediterranean environments and, in particular, that this trait could be beneficial only in germplasm with indeterminate growth habit. The advantage of dwarfism in this germplasm was larger in the cropping seasons that led to greater vegetative development, when the competition between vegetative and reproductive sinks was accentuated. In contrast, dwarfism resulted in a yield disadvantage for semi-determinate material, possibly because of constraints that it introduced in the plant's source ability, particularly that for light interception (as suggested by the marked reduction in the crop aerial biomass). However, trait assessment on a larger genetic base would be necessary, to confirm the dependency on growth habit of the value of dwarfism for white lupin targeted to Mediterranean environments.

Acknowledgments. This study was funded by the Italian Ministry of Agriculture, Food and Forestry Policy within the project 'Plant Genetic Resources – FAO Treaty (RGV-FAO)'. We are grateful to Michele Lilliu and Alessandro Longu for their technical assistance.

References

Annicchiarico P (2008) Adaptation of cool-season grain legume species across climatically-contrasting environments of southern Europe. Agron J 100:1647–1654

Annicchiarico P, Carroni AM (2009) Diversity of white and narrow-leafed lupin genotype adaptive response across climatically-contrasting Italian environments and implications for selection. Euphytica 166:71–81

Annicchiarico P, Harzic N, Carroni AM (2010) Adaptation, diversity, and exploitation of global white lupin (*Lupinus albus* L.) landrace genetic resources. Field Crops Res 119:114–124

Arnoldi A (2005) Optimized processes for preparing healthy and added value food ingredients from lupin kernel, the European protein-rich grain legume. Aracne, Rome

Boschin G, D'Agostina A, Annicchiarico P, Arnoldi A (2007) The fatty acid composition of the oil from *Lupinus albus* cv. *Luxe* as affected by environmental and agricultural factors. Eur Food Res Technol 225:769–776

Harzic N, Huyghe C (1996) Dwarfism does not modify mean area per leaf and light interception in indeterminate autumn-sown white lupin. J Agric Sci 127:337–345

Harzic N, Huyghe C, Papineau J (1995) Dry matter accumulation and seed yield of dwarf autumn-sown white lupin (*Lupinus albus* L.). Can J Plant Sci 75:549–555

Harzic N, Huyghe C, Papineau J, Billot C, Esnault R, Deroo C (1996) Genotypic variation of seed yield and architectural traits in dwarf autumn-sown white lupin. Agronomie 16:309–319

Julier B, Huyghe C, Papineau J, Milford GFJ, Day JM, Billot C, Mangin P (1993) Seed yield and yield stability of determinate and indeterminate autumn-sown white lupins (*Lupinus albus*) grown at different locations in France and the UK. J Agric Sci 121:177–186

Noffsinger SL, Huyghe C, van Santen E (2000) Analysis of grain-yield components and inflorescence levels in winter-type white Lupin. Agron J 92:1195–1202

Deciphering Grain Legumes Quality Riddle: The Genomics of Bioactive Compounds

M. C. Vaz Patto[1]([✉]), E. Mecha[1], A. B. Pereira[2], S. T. Leitão[1],
M. L. Alves[1], and M. R. Bronze[1,2,3]

[1] ITQB NOVA, Instituto de Tecnologia Química e Biológica António Xavier,
Universidade Nova de Lisboa, Oeiras, Portugal
cpatto@itqb.unl.pt
[2] iBET, Instituto de Biologia Experimental e Tecnologica, Oeiras, Portugal
[3] iMED, Faculdade de Farmácia, Universidade de Lisboa, Lisbon, Portugal

Abstract. Grain legumes are a vital source of inexpensive plant-based proteins and amino acids and, eaten as part of a healthy diet, they help to prevent chronic diseases. Potential health benefits of grain legumes have been mainly attributed to the presence of secondary metabolites with antioxidant activity. Breeding for improved end-user's quality is a complex task due to trait interaction. Some of these metabolites, such as phenolic compounds, that act as health-promoting agents are also considered antinutrients, influencing both taste and consumers' acceptability. Under the scope of LEGATO and BEGEQA projects, we studied the genetic architecture of the total phenolic compounds content in grass pea and common bean seeds. Association mapping approaches were applied to both common bean and grass pea germplasm collections and the genomic regions controlling these bioactive compounds concentration were identified. This work will contribute to the development of grain legume varieties with better nutritional quality, meeting consumers' expectations at the same time that support the improvement and production of a plant resource with a vital role in sustainable agriculture.

Keywords: Grain legumes · Bioactive compounds · GWAS

1 Introduction

Worldwide, consumers are increasingly concerned with food quality and environmental sustainability. Grain legumes answer to these concerns due to their rich nutritional composition, health promoting components and functional food status.

Yet the nonalignment of breeding objectives and end-users preferences is hampering their use in Europe. Breeding for improved end-user's quality is a complex task due to trait interaction. Particular metabolites act as antinutrients as well as health-promoting agents, influencing both taste and consumers' acceptability. This is the case of phenolic compounds. Phenolic compounds have been extensively studied due to their diverse health benefits as antioxidants, and for preventing chronic inflammation, cardiovascular diseases, cancer and diabetes (Acosta-Estrada et al. 2014). Nevertheless phenolic compounds might also contribute to increase astringency of food products (Cheynier 2012)

G. Brazauskas et al. (Eds.): *Breeding Grasses and Protein Crops in the Era of Genomics*, pp. 118–120, 2018.
https://doi.org/10.1007/978-3-319-89578-9_21

and reduce bioaccessibility of some nutrients, such as the micronutrients zinc and iron (La Frano et al. 2014).

Under the scope of LEGATO and BEGEQA projects, we are presently interested to unravel the genetic basis of the level of several of these dual action bioactive compounds. The elucidation of their complex genetic control will allow the development of fast and reliable screening methods for the most relevant consumer quality traits.

The aim of the current study was to identify the genomic regions controlling total phenolic content in grass pea and common bean in order to pave the way for grain legume nutritional quality improvement.

2   Material and Methods

To reach the study goals, 100 grass pea (*Lathyrus sativus* L.) and 100 common bean (*Phaseolus vulgaris* L.) accessions were field trialled. The grass pea trial was established in Córdoba, Spain, and the common bean field trial was established in Sintra, Portugal, using a randomized complete block design with two repetitions, during one growing season.

Dry mature seeds of common bean and grass pea were milled (Falling n° 3100 – Perten, Sweeden) to a particle size of 0.8 mm and stored at −20 °C until analysis. Phenolic compounds were extracted with 50:50 ethanol:water and the total phenolic content was determined by the Folin-Ciocalteu spectrophotometric method as described in Stamatakis et al. (2009) and expressed as gallic acid equivalents/100 g sample.

The same collections of accessions were screened using DArT-seq based SNP markers.

To detect significant associations between the grain legumes total phenolic content and the SNP molecular markers, a genome-wide association study (GWAS), joining the phenotypic information from the metabolite profile from one single environmental trial per grain legume with the genotypic information, was performed (GenStat, 18th Ed.). A mixed linear model analysis, including population structure and familial relatedness, was used to identify significant associations.

3   Results and Discussion

The high genetic diversity of common bean and grass pea accessions under study was reflected on a high diversity in the total phenolic content.

The average value observed for grass pea samples was 68 mg gallic acid equivalents/100 g sample, the minimum value was 46 mg gallic acid equivalents/100 g sample, and the maximum value was 102 mg gallic acid equivalents/100 g sample. In the case of the common bean samples, average value was 196 mg gallic acid equivalents/100 g sample, minimum value was 54 and maximum value 349 mg gallic acid equivalents/100 g sample.

The detected interesting sources of phenolic compounds may be used in future breeding programs to improve grain legume quality, with a potential for the development of broad and locally adapted high phenolic content varieties.

After genotypic data quality control, converting the residual heterozygous SNP markers to missing data, removing markers with more than 25% missing data, and removing markers with minor allele frequency smaller than 5%, only 879 grass pea SNPs and 12448 common bean SNPs were used in the following analysis.

As implemented in GenStat software, population structure using the Eigen analysis, and familial relatedness using a kinship matrix based on the Dice coefficient were assessed. A mixed linear model analysis, including population structure and familial relatedness, was used to identify significant associations. According to the inflation factor value, and the inspection of the quantile-quantile (Q-Q) plots of the $-\log 10$ (P-values), the best relationship model for the total phenolic content was the model accounting for familial relatedness using the kinship matrix in grass pea, and in common bean the model accounting for population structure using the Eigen analysis.

Ten significant marker-trait associations were detected for common bean total phenolic content (in chromosomes 1, 5 and 7) and one significant marker-trait association was detected for grass pea total phenolic content. Since there is no genome sequence available for grass pea this marker could not be located yet on a particular grass pea chromosome.

4 Conclusion

Several regions with significant marker-trait associations for total phenolic content were detected in grass pea and common bean.

Presently a second year of field evaluation is going on for both collections of accessions. This will be fundamental to better estimate the environmental effects and to study the genotype x environmental interaction on the total phenolic content of both grain legumes. A more detailed characterization on the individual phenolic compounds is also ongoing. If after the multi-environment data analysis these marker-trait associations are still detected, functional markers will be developed based on these markers to redirect breeding towards more attractive grain legume varieties.

Acknowledgments. FCT Portugal for BEGEQA project (PTDC/AGR-TEC/3555/2012); PhD fellowships SFRH/BD/89287/2012 and SFRH/BD/92160/2013, FCT Investigator Program (IF/01337/2014); R&D unit, UID/Multi/04551/2013. ESAC-IPC and Germplasm bank in the Research Unit of Biotechnology and Genetic Resources, INIAV, Portugal, for common bean samples. CRF-INIA, Spain and USDA-ARS for grasspea samples.

References

Acosta-Estrada BA, Gutiérrez-Uribe JA, Serna-Saldívar SO (2014) Bound phenolics in foods, a review. Food Chem 152:46–55

Cheynier V (2012) Phenolic compounds: from plants to foods. Phytochem Rev 11:153–177

La Frano MR, de Moura FF, Boy E, Lonnerdal B, Burri BJ (2014) Bioavailability of iron, zinc, and provitamin A carotenoids in biofortified staple crops. Nutr Rev 72:289–307

Stamatakis G, Tsantila N, Samiotaki M, Panayotou GN, Dimopoulos AC, Halvadakis CP, Demopoulos CA (2009) Detection and isolation of antiatherogenic and antioxidant substances present in olive mill wastes by a novel filtration system. J Agric Food Chem 57:10554–10564

Manipulation of Rumen Nitrogen Flows by Use of Festulolium Hybrids in Rumen Simulation Fermenters

S. Kamau, A. Belanche, M. Humphreys, and A. Kingston-Smith(✉)

Institute of Biological Environmental and Rural Sciences,
Aberystwyth University, Aberystwyth, UK
ahk@aber.ac.uk

Abstract. *Festulolium* is an interspecific hybrid between *Lolium* and *Festuca* species which combines the superior forage quality characteristics of ryegrass with fescue's ability to grow in harsh environments. The aim of this study was to perform an integrated study of rumen nitrogen metabolism by using rumen simulation to explore the extent to which swards containing *Festulolium* populations previously shown to have decreased rates of endogenous protein degradation can confer advantageous protein utilisation in comparison with a currently market leading diploid ryegrass. An *in vitro* experiment was therefore conducted using three *Festulolium* hybrids: *L. perenne* × *F. glaucescens* (LpFg); *L. perenne* × *F. mairei*(LpFm); and *L. multiflorum* × *F. glaucescens* (LmFg) and *Lolium perenne* (Lp) as a control. Festulolium hybrids had higher conversion ratios of nitrogen into microbial protein than *Lolium perenne* thus suggesting greater potential to improve ruminal nitrogen use efficiency in a fresh feeding system. This would help to increase production efficiency as well as decreasing the environmental impact of livestock agriculture.

Keywords: *Lolium perenne* (perennial ryegrass) · *Festulolium* hybrids
Rumen · Nitrogen · Protein

1 Introduction

Inefficient capture of forage nitrogen by ruminants is a significant problem for livestock farming which can lead to environmental problems such soil eutrophication and ground-water pollution. Briefly, excessive proteolysis in the rumen means that dietary protein is broken-down into amino acids and ammonia and can exceed the microbial capacity to incorporate those N sources into their microbial protein which should represent the main amino-acid source for the ruminant. As a result, the excess ammonia, which can represent up to 70% of ingested nitrogen, is excreted by the animal mainly as urinary nitrogen (MacRae and Ulyatt 1974). In consequence high production animals receive supplementation with rumen by-pass protein, often in the form of soya due to its low degradability in the rumen, representing an increased cost of production, with significant amounts of nitrogenous pollutants released to the land and water. We have shown previously in *in vitro* and *in sacco* studies that endogenous plant responses to rumen conditions contribute to inefficient protein use, and hence

© Springer International Publishing AG, part of Springer Nature 2018
G. Brazauskas et al. (Eds.): *Breeding Grasses and Protein Crops in the Era of Genomics*, pp. 121–124, 2018.
https://doi.org/10.1007/978-3-319-89578-9_22

forage-based solutions to this problem based on genetic improvement of the feed are possible (Kingston-Smith et al. 2013). *Lolium perenne* (perennial ryegrass) is a commonly used forage grass that would benefit from enhanced post-ingestion protein stability. One technique that has shown promise is to combine *Lolium* with *Festuca* genomes to generate *Festulolium* hybrids. Here we have used continuous flow rumen simulation to explore the extent to which swards containing *Festulolium* populations previously shown to have decreased rates of endogenous protein degradation can confer advantageous protein utilisation in comparison with a currently market leading diploid ryegrass (Humphreys et al. 2014).

2   Material and Methods

Four forage genotypes; Lp (*Lolium perenne* var AberMagic), and three *Festulolium* hybrids: LpxFg (*Lolium perenne × Festuca glaucescens*), LpxFm (*Lolium perenne × Festuca mairei*) and LmxFg (*Lolium multiflorum × Festuca glaucescens*) were randomly allocated to the fermenter vessels contained in 16 identical Rusitec (continuous flow rumen simulation technique) vessels of 900 mL volume (4 per treatment). Vessels were inoculated with rumen fluid obtained from 4 rumen-cannulated cows and the trial was conducted during a 12 day period. All forages were cut into 1 cm lengths before being weighed into nylon bags, introduced into the fermenters where they remained for 48 h. The liquid dilution rate was maintained at 3.65%/h by continuous infusion of artificial saliva. After 9 days adaptation 2.3 mg ^{15}N/d was continuously infused to label the microbial protein. Days 10 to 12 of the experiment were used to determine the diet degradability, gas production and rumen fermentation, while effluent was collected during days 13 and 14 and mixed with their correspondent bag residue to reconstitute the digest a flow. Finally, on day 15 liquid- (LAB) and solid-associated bacteria (SAB) were isolated to determine microbial synthesis as previously described (Belanche et al. 2013). Data were analysed by ANOVA blocking by cow inoculum.

3   Results and Discussion

There was no detectable difference in the total volatile fatty acid (VFA) production between genotypes suggesting a similar rumen degradability across genotypes. A significant difference was detected in the total nitrogen production fermentation process. LpxFg produced the lowest amount of total nitrogen as compared to the other three genotypes, which had relatively similar production. The amount of non-ammonium nitrogen (NAN) produced was similar across all genotypes. Significant differences were detected in the production of ammonia. Fermentations with LmxFg and Lp had the highest amounts of ammonia while LpxFg recorded the lowest suggesting a lower protein degradation for this latter genotype. No substantial differences among the four forages were observed in the overall proportion of both liquid and solid associated nitrogen (LAB-N and SAB-N respectively) derived from ammonia suggesting that ammonia availability did not limit microbial growth under any dietary situation (Table 1).

Table 1. Rumen function

Rumen parameters	Lp	LpxFg	LpxFm	LmxFg	SED	p-value
Total VFA (mmol/d)	23.6	20.5	26.2	28.9	3.32	ns
Total N (mg/d)	168.3[a]	144.1[b]	167.5[a]	169.8[a]	7.3	*
Total Non-ammonia N (mg/d)	119.2	107.4	125.8	118.6	7.82	ns
Total Ammonia N (mg/d)	49.1[a]	36.7[b]	41.7[ab]	51.2[a]	4.46	*
Total Microbial N-LAB (mg/d)	70.2[bc]	63.1[c]	76.9[ab]	85.8[a]	5.01	**
Total Microbial N-SAB (mg/d)	68.8	65.8	80.9	81.6	7.65	ns
Total Microbial N-LAB from NH3	0.49	0.48	0.49	0.48	0.0285	ns
Total Microbial N-SAB from NH3	0.50	0.46	0.47	0.51	0.0628	ns
Total Gas produced (L/d)	1.89	1.69	2.04	1.85	0.224	ns
Methane (mL/d)	83.7	67.4	85.7	81.5	20.01	ns

$**P < 0.01$, $*P < 0.05$, ns not significant ($n = 4$).

There were differences in the microbial protein synthesis across genotypes. Considering the liquid associated bacteria as reference, *Festulolium* hybrids LmxFg and LpxFm had +22,2% and +9,5% greater microbial protein synthesis than control ($P = 0.0089$) indicating more favourable rumen conditions for microbial growth (Fig. 1). Based on these observations, two out of the three *Festulolium* hybrids tested (LmxFg and LpxFm) were suggested to have the greater potential in terms of better nutrient-use efficiency and would be recommended onto the list of forages of choice in ruminant production.

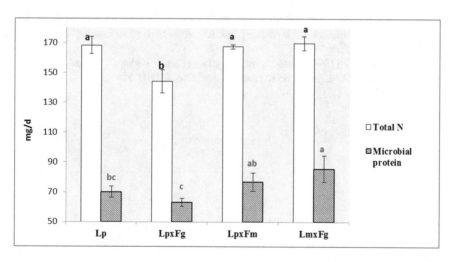

Fig. 1. Nitrogen flows during fermentation of forage genotypes in rumen simulated technique. Mean values of samples recovered after 24 h of forage administration

4   Conclusion

Inefficient capture of feed nitrogen by production ruminants is a major problem for both farm economics and the environment. In addition, increasing consumer concern about traceability and sustainability are driving efforts to increase use of farm grown rather than imported feeds. This study indicates the potential of Festulolium hybrids to provide an improved feed for grazing cattle. Encouragingly, the *Festulolium* hybrids showed potential for improved nitrogen partitioning during fermentation of fresh forage compared with a commonly used ryegrass. In practical terms this means that by use of a Festulolium based sward, a greater proportion of feed nitrogen would be retained by the animal for milk or meat production and nitrogen load to the land would be decreased in comparison to grazing of a ryegrass sward. Further work is underway to determine the basis for this differential and the extent of genetic control of the trait.

Acknowledgments. This research work was funded by Aberystwyth University DCDS (SK), BBSRC grant BBS/E/W/10964A01 (AB, TED, AK-S).

References

Belanche A, Lee MRF, Moorby JM, Newbold CJ (2013) Comparison of ryegrass and red clover on the fermentation pattern, microbial community and efficiency of diet utilisation in the rumen simulation technique (Rusitec). Anim Prod Sci 53:1052–1064

Humphreys MW, O' Donovan SA, Farrell SM, Gay AP, Kingston-Smith AH (2014) The potential of novel *Festulolium* (2n = 4x = 28) hybrids as productive, nutrient-use-efficient fodder for ruminants. Food Energy Secur 3(2):98–110

Kingston-Smith AH, Marshall AH, Moorby JM (2013) Breeding for genetic improvement of forage plants in relation to increasing animal production with reduced environmental footprint. Animal 7:79–88

MacRae J, Ulyatt MJ (1974) Quantitative digestion of fresh herbage by sheep. 2. Sites of digestion of some nitrogenous constituents. J Agric Sci 82:309–319

Paternity Test for Forage Yield Improvement in Timothy Polycross Breeding

T. Tanaka[1]([✉]), K. Tamura[2], K. Ashikaga[1], H. Fujii[1], and T. Yamada[3]

[1] Hokkaido Research Organization, Kitami Agricultural Experiment Station,
Kunneppu, Hokkaido, Japan
tanaka-tsuneki@hro.or.jp
[2] Hokkaido Agricultural Research Center, NARO, Sapporo, Hokkaido, Japan
[3] Field Science Center for Northern Biosphere,
Hokkaido University, Sapporo, Hokkaido, Japan

Abstract. Paternal identity knowledge can enhance selection gains in polycross breeding of forage grasses. This study assessed paternal parents selected through timothy (*Phleum pratense* L.) breeding without paternal information in Japan. Paternal assignment was conducted based on a simple exclusion way using 27 genomic-SSR markers. General combining ability (GCA) for forage yield was examined based on polycross progeny tests. Results showed an unbalanced number of indirectly selected paternal parent and inferior paternal parent in GCA for forage yield. These results indicate that paternal selection can give continuous improvement avoiding inbreeding depression and that it can enhance selection gains for forage yields in timothy polycross breeding.

Keywords: *Phleum pratense* · Paternal selection · Forage yield

1 Introduction

Timothy (*Phleum pratense* L.) has high forage quality and good adaptability to cold and humid northern climates (Tamaki et al. 2010). Forage yield continues to be an important objective for timothy breeding in Japan. However, despite intensive breeding efforts, gains by polycross breeding have been low. Paternal selection with paternity tests using molecular markers might overcome the improvement lag in polycross breeding (Riday 2011). This study was conducted to assess indirectly selected paternal parents through polycross breeding without paternal information and to assess the possibility of paternal selection for forage yield improvement in timothy polycross breeding.

2 Material and Methods

2.1 Plant Material and General Combining Ability for Forage Yield

Two early maturing timothy polycross groups consisting of 64 parents and 80 individual progeny plants were used for this study. Polycross was conducted using 36 (set I) and 28

© Springer International Publishing AG, part of Springer Nature 2018
G. Brazauskas et al. (Eds.): *Breeding Grasses and Protein Crops in the Era of Genomics*, pp. 125–130, 2018.
https://doi.org/10.1007/978-3-319-89578-9_23

(set II) parental clones with nine replicates in 2010 at Kitami Agricultural Experiment Station (KAES; 43°47′N, 143°42′E). Selection from their polycross progenies was conducted by among- and within- half-sib family selection (AWHS) during 2011–2013, where drilled-row plots for selecting the maternal lines and spaced-plant plots for selecting individuals derived from the selected lines were settled simultaneously and separately (Tamaki et al. 2010). The polycross progeny test (PXPT) in drilled-row plots (two seedling rows with 0.85 m long and 0.25 m apart in each plot) was established in a randomized complete block design with four replicates using seeds of the 64 progenies in 20 May 2011 at KAES. Individual phenotypic selection (IPS) in spaced-plant plots (at 0.75 × 0.6 m distances) was established using 7344 seedlings of the progenies in 25 May 2011 at KAES. The selection of 19 maternal lines was carried out based on forage yields in PXPT. The selection of 80 progeny plants derived from the selected maternal lines was conducted by IPS focused on winter hardiness, plant vigor, lodging resistance, disease resistance, competitiveness toward legumes, and forage quality. Cumulated dry matter yields (Mg ha^{-1}) over two years (2012–2013) were examined in the polycross progeny test. General combining ability for forage yield (GCA) of the polycross parents was calculated based on the cumulated dry matter yields as (Griffing 1956) $GCA_i = [(p-1)/(p-2)] (X_i - \mu)$, where X_i stands for the value of the progeny i, μ is the mean of all progenies, and p signifies the parental number of the polycross.

2.2   Genomic-SSR Genotyping and Paternity Test

Twenty-seven primer pairs were chosen from genomic-SSR markers localized on the diploid timothy (*Phleum nodosum* L.) map (Cai et al. 2009). Electrophoresis was performed on GenomeLab GeXP (Beckman Coulter, Brea, CA, USA) following the manual. Fragment sizes were ascertained using default fragment analysis parameters of the auxiliary software. Allele phenotypes of the plants were scored visually using a binary code (1/0) for the presence or absence of allele peaks without knowing the allele dosage. For each progeny plant, alleles were compared with their respective maternal alleles. Alleles not observed in the mother were recorded as paternally derived. Marker-based paternity tests used the ratio of common allele number between a given progeny plant and paternal parent candidate to the number of paternally derived alleles as the index of paternal determination. The indexes of paternal candidates for a given progeny plant were analysed by Smirnov–Grubbs test for one outlier. If the highest index was a significant outlier, then the candidate was identified as a paternal parent. Goodness-of-fit of frequency distribution of the determined paternal parents to expected probabilities was tested in each polycross group by the Pearson's chi-squared test with simulated p-value on 10,000 replicates employing the function "chisq.test" in the R statistical package "stats" (GNU General Public License).

3   Results and Discussion

The genomic-SSR analysis detected 570 alleles, 62–112 alleles per genotype, and 24–59 paternally derived alleles per progeny plant. The highest indexes for paternal determination were 0.39–0.85. The value was a significant ($p < 0.05$) outlier in all cases (Figs. 1 and 2).

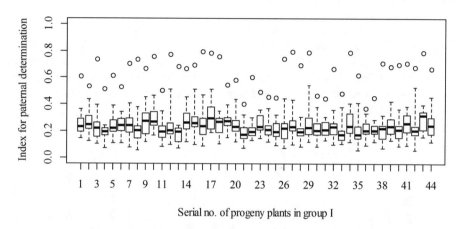

Fig. 1. Boxplots of the index for paternal determination for 44 polycross progeny plants in group I consisting of 36 parents. The index is calculated as ratio of common number between a given progeny plant and paternal parent candidate to the number of paternally derived alleles in genomic-SSR analysis. Outliers were analyzed using a Smirnov–Grubbs test.

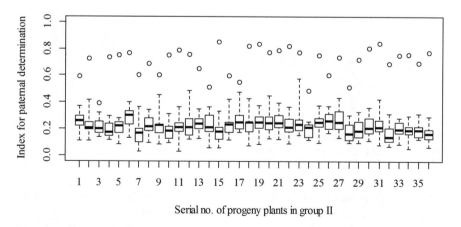

Fig. 2. Boxplots of the index for paternal determination for 36 polycross progeny plants in group II consisting of 28 parents.

Thirty-one parents were selected indirectly as paternal parents through AWHS (closed bars in Figs. 3 and 4). Chi-squared values of the goodness-of-fit test were 49.1 ($p = 0.0573$, group I) and 141.8 ($p < 0.001$, group II). Results showed clearly that the AWHS in group II caused unbalanced selection for paternal parents, although it did not occur in group I. Therefore, marker-based paternity tests were used for redressing the balance of parental number in timothy polycross breeding because AWHS might have a risk of unbalanced selection causing an inbreeding depression in recurrent selections and/or variety synthesis.

Selected maternal parents (opened bars in Figs. 3 and 4) showed clearly that the breeders had selected maternal lines based on GCA for forage yields. Meanwhile, the indirectly selected paternal parents included inferior parents in GCA for forage yields (closed bars in Figs. 3 and 4). This result indicated that marker-based paternity tests can also be used to enhance selection gain by removing the progeny plants derived from inferior parents.

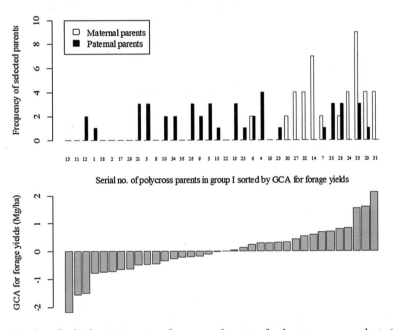

Fig. 3. Number of selections as a maternal or paternal parent of polycross progeny plants (upper part) and general combining ability (GCA) for forage yields of their parental candidates (lower part) in group I.

The reason for the unbalanced selection in group II might be IPS per se for the array of target traits, and/or unequal polycross mating described in Riday et al. (2013). Further studies must be conducted for interpretation of the occurrence of unbalanced selection in AWHS. Moreover, it is necessary to consider selection intensity in marker-based paternal selection and a sufficient number of plants before the selection for its establishment in timothy polycross breeding.

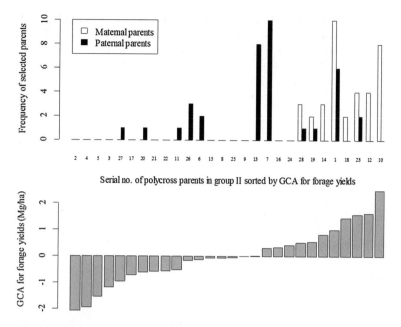

Fig. 4. Number of selections as a maternal or paternal parent of polycross progeny plants (upper part) and general combining ability (GCA) for forage yields of their parental candidates (lower part) in group II.

4   Conclusion

Marker-based paternity tests in two polycross types of timothy revealed unbalanced numbers of selected-parents and inferior paternal parents in GCA for forage yield. These results indicate the possibility of using paternal selection in timothy polycross breeding in terms of removing inferior paternal parents and redressing the imbalance of selected parents.

Acknowledgments. Molecular marker analysis in this study was supported by grants from the Northern Advancement Center for Science & Technology, and JSPS KAKENHI Grant Number 15K21609.

References

Cai HW, Inoue M, Yuyama N, Hirata M (2009) Genome mapping in cool-season forage grasses. In: Yamada T, Spangenberg G (eds) Molecular Breeding of Forage and Turf. Springer, New York, pp 173–183

Griffing B (1956) Concept of general and specific combining ability in relation to diallel crossing systems. Aust J Biol Sci 9:463–493

Riday H (2011) Paternity testing: a non-linkage based marker-assisted selection scheme for outbred forage species. Crop Sci 51:631–641

Riday H, Johnson DW, Heyduk K, Raasch JA, Darling ME, Sandman JM (2013) Paternity testing in an autotetraploid alfalfa breeding polycross. Euphytica 194:335–349

Tamaki H, Baert J, Marum P (2010) Timothy. In: Boller B, Posselt UK, Veronesi F (eds) Fodder Crops and Amenity Grasses. Handbook of Plant Breeding, vol. 5. Springer, New York, pp 324–338

Preliminary Estimation of Variation of Alkaloids Content in White Lupin (*Lupinus albus* L.) Collection

W. Rybiński[1]([⊠]), M. Kroc[1], W. Święcicki[1], P. Wilczura[1], K. Kamel[1], P. Barzyk[2], and W. Mikulski[1]

[1] Institute of Plant Genetics, Polish Academy of Sciences, Poznań, Poland
`wryb@igr.poznan.pl`
[2] Poznań Plant Breeders Ltd., Wiatrowo Plant Breeding Branch, Wągrowiec, Poland

Abstract. Grain legume crops are crucial elements of global agriculture and nutrition, both as food and feed, since they are major sources of plant protein. Of the many species of genus *Lupinus*, *L. angustifolius*, *L. luteus* and *L. albus* (characterized by high protein content) are particularly important from the agricultural point of view. Among them, white lupin (*L. albus*) has unexploited potential in Poland as a crop plant due to its high seed yield as well as protein and oil content in the seeds. In white lupin breeding, the reduction of alkaloid content in seeds is important. The paper presents an assessment of the variability in total alkaloid content and qualitative composition of alkaloids in seeds of white lupin. Plant material constituted 150 accessions from domestic collections, divided into five classes: wild collected material, landraces, induced mutants, breeding lines, and cultivars. Total alkaloid content was analysed along with qualitative composition expressed by four major alkaloids, and broad variation was found. The total alkaloid content varied from 0.008 to 11.1% of seed dry weight. An apparent decrease in alkaloid content was shown in breeding lines and cultivars, indicating a clear influence of domestication. The four major alkaloids (abundance >1%) were lupanine (mean 71.3%), 13–hydroxylupanine (10.46%), multiflorine (6.8%) and angustifoline (3.8%). The lower alkaloid content of breeding lines and cultivars, as compared to wild lines and landraces, was associated with higher relative contents of 13–hydroxylupanine and angustifoline and lower relative contents of lupanine and multiflorine.

Keywords: Alkaloids · Genetic resources · *Lupinus albus* · White lupin

1 Introduction

Lupins are economically and agriculturally valuable plants and together with other legumes represent an important source of protein in human diet. In the genus *Lupinus* more than 200 species are known, but only four (*L. angustifolius* L., *L. luteus* L., *L. albus* L. and *L. mutabilis* Sweet) are of agronomic interest. White lupin (*L. albus*) is a relatively neglected legume (including in Poland), but it has the longest history for human consumption of any lupin species, dating back to pre–Roman times (Cowling

© Springer International Publishing AG, part of Springer Nature 2018
G. Brazauskas et al. (Eds.): *Breeding Grasses and Protein Crops in the Era of Genomics*, pp. 131–136, 2018.
https://doi.org/10.1007/978-3-319-89578-9_24

et al. 1998a; Kurlovich 2002). It was the only cultivated lupin in Ancient Greece. Currently white lupin is grown around the Mediterranean and Black Seas, and in the Nile valley as far as Sudan and Ethiopia, in South Africa and in both Americas. It is characterized by such valuable traits as high protein content (32–36%) and high potential seed yield, but it remains a minor pulse crop in Poland (Rybiński et al. 2014). The main disadvantages of white lupin in Polish agriculture are its long vegetation period and susceptibility to antracnose (Święcicki et al. 2015). Human consumption of lupin has increased in recent years after the alkaloids have been removed from seeds, first by debittering and more recently by breeding to produce sweet cultivars for human food and animal feed. For white lupin as a high-protein crop, the main attention is focused on protein quality (the amino acid profile) and much less attention is given to the alkaloid content.

There are many toxic alkaloids present in the genus *Lupinus,* including pyrrolizidine and piperidine alkaloids (Panter et al. 1998). The lupin alkaloids in the species of agricultural importance are usually derivatives of quinolizidine, so they are called quinolizidine alkaloids (QAs). Over 100 QAs has been reported in the genus (Wink et al. 1995). The main role of the alkaloids is to provide the plant a chemical defence against herbivores (Wink 1992). The alkaloid fractions may have some potential as plant growth regulators, and as natural pesticides, because of their deterrent or insecticidal properties (Jimenez et al. 1982). Lupin alkaloids can affect the central nervous system of mammals (Maknickiene and Askaviciute 2008). QA intoxication is characterized by trembling, shaking, excitation and convulsion, and can lead to anticholinergic syndrome (Kurzbaum et al. 2008). Alkaloids make lupin seeds bitter and are toxic when ingested by human or animals, but bitter landraces of white lupin are grown today for consumption of seed as a snack in several Mediterranean countries. Since the development of the first sweet lupin varieties in the 1930s and 1940s, sweet lupins have slowly made inroads into modern food products including pasta, biscuits and sprouts (Cowling et al. 1998a). According to literature data, alkaloids are the major barrier for the commercial acceptance of lupin in human food and modern animal feed industries, and breeders eliminate bitter types from their material (Cowling et al. 1998b; Święcicki et al. 2015). To release new and improved sweet cultivars on the market, it is necessary to use as initial material in breeding programmes the genotypes with extremely low alkaloid content. Such possibilities may be found in genetic resources through examination of genetic diversity, geographical distribution and conservation of lupins in the Old World and South America (Hackbarth and Troll 1956; Kamel et al. 2016). This paper presents an assessment of the variability of seed alkaloid content and composition in the domestic collection of white lupin.

2   Material and Methods

A total 150 accessions of white lupin from the domestic collection were investigated. Accessions originated from countries of Europe, Asia, Africa, North America, South America and Australia.

The investigated plant material belong to five classes of origins:

CO – wild lines and primitive populations originating from places of natural distribution and collection missions;
LR – landraces;
MU – mutants;
XD – breeding lines; and
CV – present and old cultivars.

Accessions were sown for regeneration in thae field experiment in Wiatrowo, Poland (52°45′9′′N, 17°8′36′′E, altitude 86 m above mean sea level). Seed samples for analyses were collected from each plot after full maturity. Total alkaloid content and individual alkaloid composition were evaluated by gas chromatography. The extraction followed the procedure presented by Kamel et al. (2016)

3 Results and Discussion

Lupin domestication was connected with the removal of seed coat hardness and pod dehiscence, along with decreasing alkaloid content (Święcicki et al. 2000). In white lupin breeding, the reduction of alkaloid content is notable, although not as great as in narrow–leafed and yellow lupin (Święcicki et al. 2015). In the present analysis, the total alkaloid content varied from 0.008 to 11.1% of the seed dry weight (Table 1). Thus, the maximum content found in *L. albus* clearly exceeded that reported for *L. angustifolius* (Kamel et al. 2016). The average alkaloid content across the analyzed accessions was 2.5%, but there were substantial differences among the five classes (Table 1) The highest mean values were in the CO and LR accessions (4.33% and 3.73%, respectively) and the lowest for CV and MU groups (1.40% and 1.05%). According to Harrison and Williams (1982), the wild forms of *L. albus* contain up to 2.2% of quinolizidine alkaloids in the seed, whereas cultivars with the *pauper* gene carried 0.02–0.05% of alkaloid in the seed. The broadest range of alkaloid variability was in the CO class and the narrowest in the MU class. The highest number of low-alkaloid accessions was found in the CV class, undoubtedly are result of selection pressure during breeding, but one cultivar with 7% alkaloid content (probably for green manure purposes) was found. A comparison of mean values in wild and primitive (CO and LR) classes with the improved accessions (CV, XD and MU) shows evidence of breeding success in the white lupin in respect of alkaloid content. Nevertheless, it is also possible to find low alkaloid accessions in CO class and vice versa – bitter accessions in CV class. In the Australian Food Standard Code (including France, UK, and New Zealand) there is a maximum limit of 200 mg/kg of alkaloid content in lupin flours and food (Cowling et al. 1998a; Resta et al. 2008).

Quinolizidine alkaloids (QAs) occur mainly in the forms of bicyclic (e.g., lupinine), tricyclic (e.g., angustifoline) and tetracyclic (e.g., lupanine, 13–hydroxylupanine, sparteine, multiflorine and α–isolupanine) (Petterson 1998) and different profiles of alkaloids are connected with different lupin species. Within a species there are usually

Table 1. Total alkaloid content (% of the seed DW) and qualitative composition of major alkaloids (percentage of total alkaloid) in seeds of *L. albus* collection

Accession class und number	Total alkaloid content	Angustifoline	Lupanine	Multiflorine	13- hydroxy-lupanine
Wild lines	4.339	2.685	77.855	9.326	6.212
(CO) - 44	0.063	0.71	57.12	0.0	0.89
Mean	11.071	7.24	87.59	17.2	16.77
Min.					
Max.					
Landraces	3.733	3.018	74.793	7.816	7.354
(LR) - 20	0.016	1.64	43.50	0.0	3.01
Mean	5.808	4,87	87.93	17.79	17,85
Min.					
Max.					
Breeding lines	2.010	4.538	66.428	6.705	12.402
(XD) - 18	0.028	0.91	29.46	0.0	1.08
Mean	6.080	11.19	80,49	16.97	26.44
Min.					
Max.					
Mutants	1.051	4.745	71.023	4.959	12.678
(MU) - 25	0,040	2.67	58.90	0.59	6.46
Mean	4.520	8.25	80.99	13.99	29.05
Min.					
Max.					
Cultivars	1.404	4.240	66.774	5.323	13.174
(CV) - 43	0.008	1.47	28.22	0.0	3.45
Mean	7.011	11.74	84.93	14.03	36.01
Min.					
Max.					
Means for accessions	2.507	3.845	71.374	6.825	10.364

4–5 major alkaloids and several minor ones (Allen 1998). In this set of accessions, most QAs were tricyclic and tetracyclic. Alkaloid composition expressed in percentage of total alkaloid content was highly differentiated. Four major alkaloids (abundance >1%) were found: lupanine, 13–hydroxylupanine, multiflorine and angustifoline (Table 1). Lupanine dominated all classes, with an average content of 71% and range from 28 to 88%, but its content was markedly higher in CO and LR accessions than in CV and XD classes (Table 1). The second major alkaloid was 13–hydroxylupanine with average content of 12–13% for XD, MU and CV accessions in contrast to 6–7% for CO and LR classes, respectively. Accessions of CO and LR classes contained more multiflorine and less angustifoline than those of XD, MU and CV classes.

4 Conclusion

Analysis of the total alkaloid content and a qualitative composition in seeds of domestic collection of *L. albus* show a very broad variation and strong differentiation of gathered accessions. Accessions the most suitable for breeding (with the lowest alkaloid content) will be selected from the whole white lupin collection analyzed (over 350 accessions).

Comparison of results obtained in individual classes of origin showed a strong influence of domestication and breeding expressed as decreased content of alkaloids.

Accessions of breeding lines and cultivar classes (characterized by low content of total alkaloid content) showed different alkaloid compositions as compared with accessions of wild and landrace classes.

References

Allen JG (1998) Toxins and lupinosis. In: Gladstones JS, Atkins CA, Hamblin J (eds) Lupin as a Crop Plant: Biology, Production and Utilization, Chap 14. CAB International Wallingford, UK, pp 411–435

Cowling WA, Buirchell BJ, Tapia ME (1998a). Lupin, Lupinus L. Promoting the conservation and use of underutilized and neglected crops. 23. Institute of Plant Genetics and Crop Plant Research, Gatersleben/International Plant Genetics Resources Institute, Rome, Italy

Cowling WA, Huyghe C, Święcicki W (1998b). Lupin breeding. In: Gladstones JS, Atkins CA, Hamblin J (eds) Lupin as a Crop Plants: Biology, Production and Utilization. CAB International Wallingford, UK pp 93–120

Hackbarth J, Troll HJ (1956) Lupins as a grain legumes and fodder plants. In: Kappert H, Rudorf W (eds) Handbook of Plant Breeding, Part IV, 2nd edn. Verlag Paul Parey, Berlin and Hamburg, pp 1–51

Harrison JM, Williams W (1982) Genetical control of alkaloids in *Lupinus albus*. Euphytica 31:357–364

Jimenez SA, Troncoso A, Muniz A, Gonzales C (1982) Pesticidas del *Lupinus mutabilis* (tarhuti). UNSAAC–NUFFIC, Cusco, Peru

Kamel KA, Święcicki W, Kaczmarek Z, Barzyk P (2016) Quantitative and qualitative content of alkaloids in seeds of narrow–leafed lupin (*Lupinus angustifolius* L.) collection. Genet Resour Crop Evol 63:711–719

Kurlovich BS (2002) Lupins (Geography, classification, genetic resources and breeding). OY International North Express, St. Petersburg, Russia–Pellosniemi, Finland, 468 p

Kurzbaum A, Safori G, Monir M, Simsolo C (2008) Anticholinergic syndrome in response to lupin seed toxity. Isr J Emerg Med 8:20–22

Maknickiene Z, Askaviciute R (2008) Alkaloid content in lupin (Lupinus L.) genotypes and vegetation periods. Biologija 54:112–115

Panter KE, Gardner RE, Shea RE, Molyneux RJ, James LF (1998) Toxic and teratogenic piperidine alkaloids from *Lupinus, Conium* and *Nicotiana* species. In: Garland T, Barr CA (eds) Toxicants and Other Natural Toxicants. CAB International, UK, pp 345–350

Petterson DS (1998) Composition and food uses of lupins. In: Gladstones JS, Atkins CA, Hamblin J (eds) Lupin as a Crop Plants: Biology, Production and Utilization, Chap 12. CAB International Wallingford, UK, pp 353–384

Resta D, Boschin G, D'Agostina A, Arnoldi A (2008) Quantification of quinolizidine alkaloids in lupin seeds, lupin based ingredients and food. In: Lupin for Health and Wealth. Proceedings of the 12th International Lupin Association, Canterbury, New Zealand, vol 1, pp 533–535

Rybiński W, Rusinek R, Szot B, Bocianowski J, Starzycki M (2014) Analysis of interspecies physicochemical variation of grain legumes seeds. Int Agrophysics 28:491–500

Święcicki W, Rybczyński J, Święcicki WK (2000) Domestication and genetics of the yellow lupin (*Lupinus luteus* L.) and the biotechnological improvements of lupins. J Appl Genet 41:11–34

Święcicki W, Kroc M, Kamel KA (2015) Lupins. In: De Ron AM (ed) Grain Legumes. Handbook of Plant Breeding, vol 10. Springer, pp 170–218

Wink M (1992) The role of quinolizidine alkaloids in plant insect interactions. In: Bernays EA (ed) Insect–Plant Interaction, vol IV. CRC Press, Boca Raton, pp 133–169

Wink M, Meiâner C, Witte L (1995) Patterns of quinolizidine alkaloids in 56 species of the genus Lupinus. Phytochemistry 38:139–153

Yields and Quality of Some Perennial Legumes in the Czech Republic

D. Knotová[1(✉)], J. Pelikán[1], J. Skládanka[2], and P. Knot[2]

[1] Agricultural Research, Ltd., Troubsko, Czech Republic
knotova@vupt.cz
[2] Department of Animal Nutrition and Forage Production,
Mendel University in Brno, Brno, Czech Republic

Abstract. In 2013 an experiment with perennial fodder crops (alfalfa, red clover 2n and 4n, interspecific hybrid between *Trifolium pratense* and *Trifolium medium* Pramedi, Hungarian clover and goat's rue) was established. The evaluation took place in 2014 and 2015 at two locations. The yields of green mass, hay, number of cuts and quality were compared.

Keywords: *Medicago sativa* · Hybrid Pramedi · *Trifolium pratense* · *Trifolium pannonicum* · *Galega orientalis* · Forage yields · Quality

1 Introduction

Perennial fodder crops of the family *Fabaceae* are an important source of fodder for livestock. In Central Europe alfalfa (*Medicago sativa* L.) and red clover (*Trifolium pratense* L.) are the dominant crops. These perennial legumes have a high quality crop production and are bred and used to a large extent in the Czech Republic. However, climate change will lead to a search of new species which will be due to higher temperatures show a rather subtropical character. Within the project: "Development and optimization methods for the determination of biogenic amines in response to increasing health security of silage" were tested and evaluated the following species: two Czech varieties of *Medicago sativa* (Holyna and Tereza), two varieties of *Trifolium pratense* (Spurt and Amos), interspecific hybrid between *Trifolium pratense* and *Trifolium medium* Pramedi, *Trifolim pannonicum* (variety Panon) and *Galega orientalis* (variety Gale).

1.1 Introducting of Tested Species

Medicago sativa is a perennial herb with a massive root system. It provides quality protein fodder with a high mineral content, especially calcium. In young state it is a suitable forage crop for all kinds of livestock. In the Mediterranean is an important fodder crop for the production of protein feed. It has a high feeding value, especially high protein content, providing several cuts in the year and it is frost resistant. The height of yields depends on genotype, environmental conditions, biological factors and agro-technical interventions (Julier 1996). In the Czech Republic 18 varieties are

© Springer International Publishing AG, part of Springer Nature 2018
G. Brazauskas et al. (Eds.): *Breeding Grasses and Protein Crops in the Era of Genomics*, pp. 137–141, 2018.
https://doi.org/10.1007/978-3-319-89578-9_25

registred of which 13 varieties have been bred in the Czech Republic. The experiment tested Czech varieties Holyna and Tereza.

Trifolium pratense is a perennial plant that has been cultivated as a crop in Andalusia in the 13th century. It is one of the basic forages, rich in protein. Compared to alfalfa, the red clover lignificates slowly, it is perfect for green fodder and is suitable for the less fertile land where alfalfa does not grow. The red clover is grown both in pure stands and in mixtures with grasses. It is used for hay, silage or grazing (Šmahel et al. 2013). Over 40 varieties are registered in the Czech Republic. The experiment tested Czech diploid variety Spurt and tetraploid variety Amos.

Variety Pramedi was bred by the way of interspecific hybridization of the *Trifolium pratense* L. cv. Tatra (2n = 4x = 28) and the *Trifolium medium* L. (2n = 8x = 64). Significant differences among hybrids and both parents were detected. The number of stems per plant is higher in this hybrid. Pramedi has short underground rhizomes which are not present in *T. pratense*. This character was inherited from *T. medium*. Pramedi is suitable as replacement for red clover in pure stand. The plant is erect, its flowering time and quality are comparable to the tetraploid varieties of the red clover. Plants have higher persistence compared to the red clover.

Trifolim pannonicum is a perennial species which is widespread in Hungary, Ukraine, Italy, in the Balkan and also in the Czech Republic. It occurs in the steppes and meadows. It has a deep taproot and short creeping rhizomes. In the older literature it was indicated that it is a forage crop with good quality (Pelikán et al. 2016). At present only one variety is legally protected in the Czech Republic (Panon).

Galega orientalis is a perennial species which is native in the Caucasus region and Armenia. Its cultivation is widespread in Baltic countries, Scandinavia, northwest Russia and Finland. In the Czech Republic it can be found in nature. It grows on moist soils rich in nutrients, optimum soils are loamy to clayey, with a pH value of 6.0–7.5. The plant has a slow initial growth, reaches a peak after five to seven years, and then gradually subsides. Fodder is suitable for livestock in the form of silage, hay or meal. The crop is not suitable for grazing. No variety is registered in the Czech Republic. The experiment tested Estonian variety Gale (Pelikán et al. 2012).

2 Material and Methods

The experiment was established on two study sites (Troubsko and Vatín) in 2013. Locality Troubsko belongs to warm and dry growing areas with an altitude of 270 m a. s.l. The average annual rainfall is 512 mm and the average annual temperature is 9.4 °C. The locality Troubsko is suitable area for growing alfalfa, but less suistable for the cultivation of red clover. Vatín research station is located at an altitude of 560 m a. s.l., the average annual rainfall is 617 mm and the average annual temperature is 6.9 °C. The experimental design was a randomized block design with four replications, plot size was 10 m^2. The yields of green mass were monitored. Dry matter samples were taken for chemical analyses and determining hay yields. All species were harvested at optimum maturity stage (butonisation). We evaluated two harvest years (2014 and 2015). Chemical analyses are available only from the first cut from the locality Troubsko. The results are evaluated by the analysis of variance.

3   Results and Discussion

In Tables 1 and 2 are yields of dry matter. Table 3 shows the quality of forage from the first cut of the site Troubsko. In 2014, the influence of the locality was reflected in the yields. In this year excellent yields in all crops were monitored in Vatín. Yields of green mass depend on the environment, the density of plants and genotype (Lamb et al. 2003). The main yield factor is the amount of usable water, either from rainfall or accessible groundwater (Lang 2011). The year 2015 was favourable for alfalfa in Troubsko, while other crops achieved higher yields in Vatín. The growing of goat's rue is problematic in Czech conditions, because *Rhizobium galegae* bacteria are absent (Usťak and Váňa 2012). Unfortunately, before sowing the seeds were not inoculated by microbiological preparations.

Table 1. Hay yield of different leguminous forage crops in Troubsko (t.ha^{-1}) ($\alpha = 0,05$)

Tested species		2014					2015				
		1st. cut (t.ha-1)	2nd. cut (t.ha-1)	3rd. cut (t.ha-1)	4th. cut (t.ha-1)	Total yield (t.ha-1)	1st. cut (t.ha-1)	2nd. cut (t.ha-1)	3rd. cut (t.ha-1)	4th. cut (t.ha-1)	Total yield (t.ha-1)
Medicago sativa (Holyna)		7.02[a]	5.07[a]	5.96[a]	2.20[a]	20.30[a]	6.84[a]	5.65[a]	2.90[a]	0.65[a]	16.05[a]
Medicago sativa (Tereza)		6.59[ab]	4.44[a]	5.87[a]	1.91[b]	18.80[a]	6.75[a]	5.33[a]	2.76[a]	0.66[a]	15.70[a]
Trifolium pratense - 2n (Spurt)		6.68[ab]	2.49[bc]	4.12[b]	0.91[c]	14.20[b]	4.27[b]	1.86[b]			6.51[b]
Trifolium pratense - 4n (Amos)		5.70[c]	3.24[b]	2.88[c]	0.37[d]	12.19[c]	3.73[b]	0.99[c]			4.68[c]
Interspecies hybrid (Pramedi)		4.85[d]	1.99[c]	2.16[c]	0.11[d]	9.12[d]	3.56[b]	1.65[b]			4.83[c]
Trifolium pannonicum (Panon)		0.78[e]	1.70[c]			2.48[e]	4.40[b]				4.40[c]
Galega officinalis (Gale)		0.45[e]				0.45[f]	2.10[c]				2.10[d]
D$_T$	0.05	0.81	1.18	0.96	0.28	1.78	1.31	0.42	0.67	0.18	1.43
	0.01	1.2	1.78	1.46	0.43	2.63	1.93	0.64	1.59	0.42	2.11

D$_T$ = Minimal significant difference
a, b, c, d, e, f = in the same column indicate significant differences (p < 0,05)

Table 2. Hay yield of different leguminous forage crops in Vatín (t.ha^{-1}) ($\alpha = 0,05$)

Tested species		2014					2015				
		1st. cut (t.ha-1)	2nd. cut (t.ha-1)	3rd. cut (t.ha-1)	4th. cut (t.ha-1)	Total yield (t.ha-1)	1st. cut (t.ha-1)	2nd. cut (t.ha-1)	3rd. cut (t.ha-1)	4th. cut (t.ha-1)	Total yield (t.ha-1)
Medicago sativa (Holyna)		7.60[b]	5.65[ab]	4.20[a]	2.35[a]	19.80[b]	5.98[a]	3.20[a]	1.70[a]	1.53[b]	12.40[b]
Medicago sativa (Tereza)		7.57[b]	6.13[a]	4.07[a]	2.52[a]	20.30[b]	6.40[a]	3.20[a]	1.60[a]	1.67[ab]	13.00[b]
Trifolium pratense - 2n (Spurt)		8.95[a]	5.72[ab]	4.10[a]	2.95[a]	21.72[a]	6.95[a]	2.33[b]	0.85[b]	1.85[a]	12.70[b]
Trifolium pratense - 4n (Amos)		8.63[a]	5.00[bc]	4.33[a]	1.58[b]	20.08[b]	5.72[a]	1.97[c]	0.77[b]	1.85[a]	14.80[a]
Interspecies hybrid (Pramedi)		9.00[a]	5.28[bc]	4.18[a]	1.25[b]	19.70[b]	5.90[a]	1.78[c]	0.75[b]	1.42[b]	8.70[c]
Trifolium pannonicum (Panon)		5.72[c]	4.75[c]	0.63[b]		10.60[c]					
Galega officinalis (Gale)		4.10[d]	2.70[d]	1.20[b]		7.90[d]					
D$_T$	0.05	0.92	0.73	0.61	0.62	1.34	2.23	0.3	0.27	0.3	1.51
	0.01	1.36	1.08	0.9	0.94	1.97		0.45	0.41	0.45	2.29

D$_T$ = Minimal significant difference
a, b, c, d = in the same column indicate significant differences (p < 0,05)

Table 3. Forage quality; first cut taken in Troubsko in years 2014 and 2015

Tested species	Ash		Nitrogenous substances		Fibre		ADF		NDF		Fat		Sugar		N	
	2014	2015	2014	2015	2014	2015	2014	2015	2014	2015	2014	2015	2014	2015	2014	2015
Medicago sativa (Holyna)	8.67[bc]	9.38[ab]	17.74[a]	21.25[a]	27.23[a]	22.63[a]	38.58[a]	34.15[a]	48.28[b]	44.32[b]	1.45[e]	1.81[bc]	6.82[b]	6.05[d]	2.84[a]	3.40[a]
Medicago sativa (Tereza)	8.95[b]	8.78[bc]	19.01[a]	21.14[a]	27.24[a]	22.43[a]	34.51[b]	33.96[a]	46.79[b]	42.02[bc]	2.05[bc]	1.89[bc]	10.10[b]	7.05[cd]	2.92[a]	3.38[a]
Trifolium pratense - 2n (Spurt)	7.41[d]	8.30[cd]	14.53[c]	18.07[b]	23.43[bc]	17.31[cd]	30.31[cd]	27.45[b]	41.47[de]	36.04[d]	1.77[cde]	1.81[bc]	17.18[a]	13.98[a]	2.33[b]	2.89[b]
Trifolium pratense - 4n (Amos)	6.99[e]	7.65[d]	12.96[d]	15.38[c]	24.81[b]	19.28[b]	32.01[d]	29.12[b]	44.25[c]	40.97[c]	1.82[cd]	1.63[c]	17.20[a]	13.20[a]	2.08[c]	2.46[c]
Interspecies hybrid (Pramedi)	7.94[cd]	8.25[cd]	16.00[b]	17.92[b]	20.29[d]	16.30[d]	26.70[e]	28.31[b]	39.84[e]	40.31[c]	2.27[b]	1.97[bc]	15.29[a]	10.81[b]	2.56[b]	2.87[b]
Trifolium pannonicum (Panon)	9.23[b]	10.08[a]	15.66[bc]	18.61[b]	23.65[bc]	24.07[a]	27.52[e]	27.57[b]	42.83[cd]	42.66[bc]	2.94[a]	2.09[b]	17.3 2[a]	7.24[c]	2.56[b]	2.98[b]
Galega officinalis (Gale)	12.15[a]	9.10[bc]	14.8 5[bc]	10.93[d]	22.27[c]	19.82[bc]	32.92[bc]	24.09[c]	51.73[a]	49.62[a]	0.65[f]	2.86[a]	7.29[b]	7.91[c]	2.38[b]	1.75[d]
D_T	0.05	0.88	0.62	1.34	1.2	1.97	1.99	2.31	2.83	2.39	2.9	0.32	0.43	4.81	1.07	0.23	0.19
	0.01	1.29	0.91	1.97	1.77	2.9	2.94	3.41	4.17	3.53	4.28	0.48	0.64	7.1	1.57	0.33	0.29

D_T = Minimal significant difference

a, b, c, d, e = in the same column indicate significant differences (p < 0.05)

4   Conclusion

Although alfalfa is a crop suitable to the conditions of southern Moravia, in the first harvest year it provided very high yields of biomass also in the highland (study site Vatín), because of sufficient rainfall. In today's changing climatic conditions it could be recommended to grow alfalfa also in highlands. Alfalfa is able to ensure high yields and forage of good quality for livestock also in these conditions. In the second crop year alfalfa provided similar forage yields as the red clover. Interspecific hybrid Pramedi provided lower yields compared to the red clover, but in the years with sufficient precipitation it is able to give even higher yields than the red clover. The Hungarian clover is a minor crop, which is expected to be used for extensive cultivation in drier localities. In the first harvest year, this species had a significantly higher sugar content compared to alfalfa and the Goat's rue and a lower protein content. This fact predetermines this species to better ensilability of forage.

Acknowledgments. Results presented in this study were obtained within the research project n° QJ1310100 "Development and optimization methods for the determination of biogenic amines in response to increasing health security of silage" that was financially supported by the Ministry of Agriculture of the Czech Republic.

References

Julier B (1996) Traditional seed maintenance and orogins of the French lucerne landraces. Euphitica 92:353–357

Lamb JFS, Sheaffer CC, Samac DA (2003) Population density and harvest maturity effects on leaf and stem yield in alfalfa. Agron J 95:635–641

Lang J (2011) Jetelovinotravní směsi jako stabilní pícninářský prvek v podmínkách měnícího se klimatu. Úroda, vědecká příloha, 312–315, ISSN 0139-6013

Pelikán J, Hýbl M, et al (2012) Czech plants of the family Fabaceae LINDL, Issue 1, Ing. Petr Baštan, Olomouc. ISBN 978-80-905080-2-6

Pelikán J, Knotová D, Hofbauer J (2016) The less known species of agricultural crops. Agriprint. s.r.o., Olomouc, p 272. ISBN 978-80-88000-06-8

Usťak S, Váňa V (2012) Možnosti pěstování a využití jestřabiny východní *Galega orientalis* Lamb. v podmínkách české republiky. Metodika pro praxi, VURV, ISBN 978-80-7427-100-7

Šmahel P, Knotová D, Lang J (2013) Situation in production of fodder crops in the Czech Republic and growing of alfalfa, red clover and grass/clover mixtures. J Mt Agric Balkans 16 (4):936–958 ISSN: 1311-0489

Breeding for Enhanced Stress Tolerance

Low Crown Rust Resistance in Norwegian Material of *Lolium perenne* and ×*Festulolium*

L. Østrem[1(✉)], T. Asp[2], M. Ghesquière[3], Y. Sanada[4], and O. A. Rognli[5]

[1] The Norwegian Institute of Bioeconomy Research (NIBIO), Fureneset, 6967 Hellevik i Fjaler, Norway
liv.ostrem@nibio.no
[2] Department of Molecular Biology and Genetics, Aarhus University, 4200 Slagelse, Denmark
[3] INRA/URP3F, Le Chêne, RD 150, CS 80006, 86600 Lusignan, France
[4] Hokkaido Agricultural Research Center, NARO Hitsujigaoka, Toyohira, Sapporo 062-8555, Japan
[5] Department of Plant Sciences, Faculty of Biosciences, Norwegian University of Life Sciences, 1432 Ås, Norway

Abstract. Norwegian cultivars and breeding materials of perennial ryegrass and *Festulolium* were planted at three locations in Denmark, France and Japan for testing of resistance against leaf diseases. In general, all plant materials were susceptible to crown rust. The highest incidence of rust attack occurred at the French site, which due to its climatic conditions might be the most suitable testing site for future scoring of similar plant material. Entries based on introgressed genetic materials from UK were most resistant towards crown rust. Crown rust resistance needs increased focus as a breeding objective in the Nordic region due to climate changes, which will most likely lead to increased infection of leaf diseases.

Keywords: Breeding · Climate change · Leaf diseases · Meadow fescue
Perennial ryegrass

1 Introduction

The Norwegian plant breeding company Graminor Ltd. develops cultivars of perennial ryegrass (*Lolium perenne* L.) and ×*Festulolium* (×*Festulolium* Aschers. et Graebn.) for Nordic conditions. The main breeding objective in these non-native species has been to improve abiotic stress resistance, especially winter survival to broaden the growing area towards the north and to continental regions. Resistance to biotic stresses like leaf diseases during the growing season has so far had low priority as breeding goals.

Global warming and increased precipitation in the Nordic region will cause favourable conditions for several plant pathogens (IPCC 2013). Fungal diseases like rusts and leaf spots, currently common during summer in warmer climate further south and west in Europe, will spread northwards, and most of our current cultivars will be

© Springer International Publishing AG, part of Springer Nature 2018
G. Brazauskas et al. (Eds.): *Breeding Grasses and Protein Crops in the Era of Genomics*, pp. 145–149, 2018.
https://doi.org/10.1007/978-3-319-89578-9_26

prone to attack by new diseases. These conditions are so far not predominant in the Nordic region, however, when looking ahead, it will necessitate increased efforts in breeding (Helgadóttir et al. 2016).

Here we present the results from testing susceptibility to crown rust/leaf rust *(Puccinia coronata)* using a multi-site field experiment with selected cultivars and breeding populations of perennial ryegrass and × *Festulolium* at locations with diverse climatic conditions.

2   Material and Methods

A total number of 21 breeding materials and cultivars of perennial ryegrass, hybrid ryegrass (*Lolium* × *boucheanum* Kunth), *Festulolium* and meadow fescue (*Festuca pratensis* Huds.) were established in field experiments in Denmark (Flakkebjerg, 55° 18′N 11°23′ E; 31 m. asl), France (Lusignan, 46°26′N 00°07′E; 134 m. asl), and Japan (Sapporo, 43°03′N 141°20′E, 29 m. asl) in 2012. In 2013, 30 plants in three replicates per entry were scored for percentage of diseased leaf area. Cutting and fertilization were conducted according to standard procedures at each site. Abdelhalim et al. (2016) presented a detailed description of the plant materials concerning parental origin, ploidy level, and region for adaptation. The data was analysed by a multi-factorial analysis using a General Linear Model (SAS 9.4, SAS Institute Inc. Cary. NC. USA) considering entry or parental groups and sites as fixed factors.

3   Results and Discussion

Parental Origin of the Entries. The five parental groups comprised diploid and tetraploid *Festulolium* and perennial ryegrass, and meadow fescue. One hybrid ryegrass cultivar was tested and grouped together with tetraploid perennial ryegrasses. When grouped according to parental origin the two groups of *Festulolium* were significantly less infected than the three remaining groups of perennial ryegrass and meadow fescue when averaged over all sites (F value 26.88; $P < 0.0001$) (Table 1, Figs. 1 and 2). The two groups of *Festulolium* were also significantly different, of which diploid *Festulolium* was less infected than the tetraploid group (34% vs. 44%). Two diploid *Festulolium* entries were significantly less infected compared with all entries. The least infected population was FuRs0348 (7.7%) originating from a cross between *L. multiflorum* x *F. arundinacea* . The population is based on surviving plants from the IBERS population Bx421. Another population with low infection rate was FuRs0353 (15.3%), originating from dihaploid plants made from the Polish cultivar Sulino. This cultivar has a similar genetic background as cv. Felopa (Ghesquière et al. 2010), which in this study was the least infected cultivar of the tetraploid *Festulolium* group, being significantly different both from the two diploids described above and the remaining entries. Also, population FuRs0356 of the diploid *Festulolium* group, also originating from IBERS material, revealed low infection rate. This is in contrast to the diploid *Festulolium* population FuRs0467 which was heavily infected (56.6%), originating from pure Nordic material obviously with no resistance to crown rust.

Meadow fescue was included in the test since it has been the most important *Festuca* parent in the Norwegian breeding program of *Festulolium*. The two main cultivars, adapted for southern (Fure) and northern (Norild) parts of Norway, were both heavily infected, 51% and 49%, respectively. During the last decade, increasing incidences of leaf diseases have been observed in regrowth of meadow fescue in agricultural leys, although not documented.

Environmental Effects. Incidence of crown rust was observed at all three sites, however, the degree of rust attacks were significantly different among sites (F value 453.35; *P < 0.0001*). Generally, high incidence of crown rust was observed in Lusignan (France) (64%) and in Flakkebjerg (Denmark) (55%), whereas in Sapporo (Japan) smaller attack was observed (16%) (Fig. 3).

A significant interaction between parental groups and sites (F value 17.55; *P < 0.0001*) was due to different ranking of the entries at the three sites. In Denmark, meadow fescue was the most infected group, whereas in France and Japan it was low to moderately infected. A change in infection levels was also observed for the diploid *Festulolium* entries, being the group with lowest infection rate in France and Denmark, while it was the most infected group in Japan.

Table 1. Observations of crown rust infections (% of leaf area) ± SE observed in five parental groups in three diverse environments.

Parental groups		FL-d	FL-t	PRG-d	PRG-t	MF
Test sites	Flakkebjerg_DK	37.7 ± 3.2	56.3 ± 2.8	56.7 ± 3.3	59.2 ± 2.8	83.1 ± 2.8
	Lusignan_FR	46.8 ± 3.3	60.8 ± 2.6	75.7 ± 2.8	80.3 ± 2.3	52.5 ± 4.3
	Sapporo_JP	17.8 ± 1.4	12.4 ± 1.3	16.6 ± 1.5	17.0 ± 1.4	13.4 ± 2.0

FL-d Festulolium diploid (1: FuRs0026, 2: FuRs0348, 3: FuRs0353, 4: FuRs0356, 5: FuRs0467), **FL-t** Festulolium tetraploid (6: Felopa, 7: FuRs0142, 8: FuRs0465, 9: FuRs9806, 10: FuRs9812), **PRG-d** perennial ryegrass diploid (11: Arka, 12: Fagerlin, 13: FuRa9805, 14: FuRa0575-79), **PRG-t** perennial ryegrass tetraploid (15: Figgjo, 16: Ivar, 17: LøRa9401, 18: FuRa9704, 19: hybrid ryegrass Fenre), **MF** meadow fescue (20: Fure, 21: Norild).

Breeding Priorities. The generally high susceptibility to crown rust observed at the French and Danish test sites illustrates that crown rust resistance in the Nordic breeding materials of these species needs improvement urgently. Otherwise, farmers will experience leys with inferior forage quality in the near future following climate change. Leaf diseases in the growing season should be included in breeding programs by introgressing genetic material from regions where the plant materials have been exposed to crown rust during longer periods and have developed natural resistance. When cultivars have acquired natural resistance, there are no signs of breakdown of crown rust resistance over time, as shown by the European multisite trial with cultivars tested every third year in a 12 years period (Schubiger et al. 2016).

For the Nordic region, it will be challenging to include both biotic and abiotic breeding objectives in the program of perennial ryegrass and *Festulolium*. Although warmer climate is expected in the future, winter survival will still be crucial for this

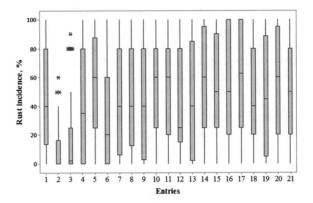

Fig. 1. Observed percentages of crown rust for the investigated entries averaged over sites. Entries 1–5 (FL-d), 6–10 (FL-t), 11–14 (PRG-d), 15–19 (PRG-t), MF (20-21) (see Table 1 for description of parental groups).

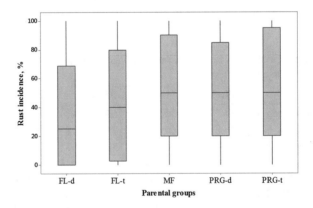

Fig. 2. Distribution of the five parental groups for crown rust infection averaged over sites (see Table 1 for description of parental groups).

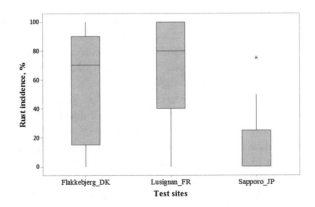

Fig. 3. Distribution of crown rust infection averaged over entries and test sites.

region. The two diploid entries FuRs0353 and FuRs0348 with high crown rust resistance in this study, demonstrated a low survival rate when tested under field conditions at Ås, Norway, due to its southern origin (Abdelhalim et al. 2016). There are, however, cultivars of perennial ryegrass available with resistance to crown rust which can be incorporated into gene pools with low resistance levels, and further improved through generations of natural selection to obtain a level of adaptation required for sufficient winter survival under Nordic conditions.

4 Conclusions

In general, low resistance to crown rust was observed across the three experimental sites in Denmark, France and Japan. The test site in France (Lusignan) was most suitable for testing crown rust resistance since the largest variation in rust attack among cultivars and breeding populations was present at this site. Thus, testing of breeding materials in similar environments should be used for future scoring of crown rust of Norwegian breeding material. Future breeding programs of perennial ryegrass and *Festulolium* will need to include genetic material with good resistance to crown rust.

Acknowledgements. The study was funded by the Norwegian Research Council and is a part of the VarClim project (199664, Understanding the genetic and physiological basis for adaptation of Norwegian perennial forage crops to future climates).

References

Abdelhalim M, Rognli OA, Hofgaard I, Østrem L, Tronsmo AM (2016) Snow mould resistance under controlled conditions and winter survival in the field in populations of perennial ryegrass, meadow fescue and Festulolium are partly dependent on ploidy level and degree of northern adaptation. Can J Plant Sci 96:579–589

Ghesquière M, Humphreys MW, Zwierzykowski Z (2010) Festulolium. In: Boller B, Posselt UK, Veronesi F (eds) Fodder crops and amenity grasses. Handbook of plant breeding, vol 5. Springer, New York, pp. 293–316. https://doi.org/10.1007/978-1-4419-0760-8_12

Helgadóttir Á, Østrem L, Collins RP, Humphreys M, Marshall A, Julier B, Gastal F, Barre P, Louarn G (2016) Breeding forages to cope with environmental challenges in the light of climate change and resource limitations. In: Roldán-Ruiz I, Baert J, Reheul D (eds) Breeding in a world of scarcity, Springer International Publishing, Switzerland, pp 3–13. ISBN 978-3-319-28930-4, ISBN 978-3-319-28932-8 (eBook)

IPCC (2013) The Physical Science Basis. Contribution of Working Group I to the Fifth Assessment Report of the Intergovernmental Panel on Climate Change. In: Stocker TF et al (eds) Cambridge University Press, Cambridge and New York, 1535p

Schubiger FX, Baert J, Ball T, et al (2016) The EUCARPIA multi-site rust evaluation − 2013 results. In: Roldán-Ruiz I, Baert J, Reheul D (eds) Breeding in a world of scarcity, Springer International Publishing, Switzerland, pp 53–62. ISBN: 978-3-319-28930-4, ISBN 978-3-319-28932-8 (eBook)

Unravelling the Genetic Control of Bacterial Wilt Resistance in Ryegrass: Achievements, Prospects and Challenges

R. Kölliker[1,2(✉)], V. Knorst[1,2], L. Hersemann[2], F. Widmer[2],
and B. Studer[1]

[1] Molecular Plant Breeding, Institute of Agricultural Sciences,
ETH Zurich, Zurich, Switzerland
roland.koelliker@usys.ethz.ch
[2] Molecular Ecology, Agroscope, Zurich, Switzerland

Abstract. *Xanthomonas translucens* pv. *graminis* (*Xtg*) causes bacterial wilt, one of the most important forage grass diseases in temperate grasslands. Molecular genetic and genomic tools have the potential to significantly benefit resistance breeding and to enable targeted resistance management. In the past, a major QTL for bacterial wilt resistance was identified in *Lolium multiflorum* and *Xtg* was shown to rely on a non-canonical type III secretion system for plant infection. Recently, a number of candidate genes for bacterial wilt resistance were identified by comparing genomic sequences of resistant and susceptible parental plants and their progeny. Comparative genomics of different *X. translucens* pathovars allowed to identify virulence traits characteristic for *Xtg*. These candidate plant resistance genes together with the bacterial virulence factors provide an invaluable resource for the development of genomics assisted selection strategies. In addition, the well characterised plant genotypes and bacterial strains serve as an ideal model system to fully understand the complex *L. multiflorum-Xtg* interaction.

Keywords: *Xanthomonas translucens* pv. *graminis* · *Lolium multiflorum*
Plant resistance · Pathogen virulence

1 Introduction

Bacterial wilt, caused by *Xanthomonas translucens* pv. *graminis* (*Xtg*), is a devastating disease on forage grasses such as *Lolium multiflorum*, leading to significant yield and quality losses (Egli et al. 1975). Although cultivars with considerable resistance to bacterial wilt have been obtained through recurrent phenotypic selection (Suter et al. 2017), marker-assisted breeding would greatly benefit efficient selection and allow to fix dominant resistant genes in this allogamous, highly heterozygous species. Tremendous technical developments in the area of DNA and RNA sequencing (Goodwin et al. 2016), together with an evolving conceptual framework on host-pathogen interactions (Pritchard and Birch 2014) have opened new avenues for characterizing the genetic control of disease resistance in plants.

© Springer International Publishing AG, part of Springer Nature 2018
G. Brazauskas et al. (Eds.): *Breeding Grasses and Protein Crops in the Era of Genomics*, pp. 150–154, 2018.
https://doi.org/10.1007/978-3-319-89578-9_27

In this paper, we summarize past achievements towards understanding the complex interaction between *L. multiflorum* and *Xtg,* highlight current prospects for developing genomic breeding tools and touch on remaining challenges regarding breeding for bacterial wilt resistance in forage grasses.

2   Past Achievements

Soon after the discovery of bacterial wilt, several pathovars of *X. translucens* have been distinguished based on their host specificity and selection for bacterial wilt resistance was soon integrated into breeding programs (Michel 2001). In a first attempt to characterize the genetic control of bacterial wilt resistance in *L. multiflorum*, Studer et al. (2006) identified a major quantitative trait locus (QTL) on linkage group (LG) 4 which explained up to 84% of the phenotypic variance. However, marker-assisted introgression of this QTL in breeding programs was hindered by the lack of sequence specific markers in the QTL region. Although major QTL are often associated with race-specific major resistance genes, a screening of 62 plant genotypes revealed no major race-specific interactions between *L. multiflorum* and *Xtg* (Wichmann et al. 2011b). Further evidence for broad-spectrum rather than race-specific resistance was gained from transcriptome analyses, where transcriptional changes usually triggered by pathogen associated molecular patterns were observed (Wichmann et al. 2011a).

However, whole genome sequencing of *Xtg* revealed the existence of a non-canonical but functional type III secretions system (T3SS; Wichmann et al. 2013). The T3SS is crucial for effector translocation from the pathogen to the host in effector-triggered immunity (Pritchard and Birch 2014). In *Xtg,* T3SS mutants have been shown to lose their ability to cause disease in *L. multiflorum* but to keep their ability for *in planta* multiplication (Wichmann et al. 2013). Another characteristic of *Xtg* strains was their relatively low genetic diversity when analysed across a larger geographic range using amplified fragment length polymorphism markers (Kölliker et al. 2006).

3   Current Prospects

In order to generate a resource for sequence-based resistance gene discovery in *L. multiflorum*, and to further characterise the QTL on LG4, the resistant and the susceptible parental plants of the *Xtg* mapping population were sequenced using Illumina HiSeq, resulting in a total of 18×10^6 heterozygous single nucleotide polymorphisms (SNPs) between the two parents (Knorst et al. 2016). In addition, genomic sequences of DNA pools of the 57 most resistant and the 50 most susceptible F_1 individuals were produced and aligned to the parental reference sequence.

Scaffolds containing such SNPs were aligned with the *L. perenne* genome sequence (Byrne et al. 2015). Candidate scaffolds allocated to LG 4 were mapped to the *L. multiflorum* linkage map using SNP data from KASP assays (LGC, Middlesex, UK). The eleven candidate scaffolds mapped to the region where the initial QTL for bacterial wilt resistance was identified, spanning a region from 54 to 74 cM on LG 4 (Fig. 1).

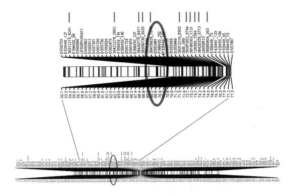

Fig. 1. Linkage group 4 of the *L. multiflorum* linkage map (improved based on Studer et al. 2006) with locations of the QTL for bacterial wilt resistance (circle) and 11 candidate scaffolds (bars) identified through genome sequencing of parental genotypes and DNA pools of resistant and susceptible progeny, respectively.

The identified candidate scaffolds not only mapped in the vicinity of the QTL, they also showed homologies to genes known to be involved in disease resistance such as for example serine/threonine kinases and therefore present a valuable resource for targeted breeding of bacterial wilt resistance.

Detailed characterisation of bacterial virulence may not only allow to better understand complex host-pathogen interactions, but also to identify and exploit plant resistant traits which correspond to bacterial virulence factors.

Comparative genomic analyses of various *X. translucens* pathovars revealed very high nucleotide identity among *Xtg* strains (>99%), high nucleotide identity among *Xtg* and other forage grass infecting pathovars (>97%) and only moderate identity among

Table 1. Average nucleotide identities of *X. translucens* pathovars pathogenic (*Xtg, Xta, Xtp*) and non-pathogenic (*Xtt, Xtc*) on forage grasses.

	Xtg29	*Xtg2*	*Xtg9*	*Xtg10*	*Xta*	*Xtp*	*Xtt*
Xtg2	**99.94**						
Xtg9	**99.93**	**99.92**					
Xtg10	**99.92**	**99.93**	**99.95**				
Xta	97.98	97.96	97.96	97.95			
Xtp	97.67	97.6	97.62	97.61	98.00		
Xtt	95.98	95.98	95.98	95.94	95.96	95.91	
Xtc	95.35	95.35	95.33	95.33	95.42	95.31	95.34

Data represent the mean percentage of identity of orthologues shared by the strains. *Xtg: X. translucens* pv. *graminis* strains from Switzerland, *Xta: X. translucens* pv. *arrhenateri* (LMG 727), *Xtp: X. translucens* pv. *poae* (LMG 727), *Xtt: X. translucens* pv. *translucens* (DSM 18974), *Xtc: X. translucens* pv. *cerealis* (CFBP 2541). Modified from Hersemann et al. (2017).

strains pathogenic and non-pathogenic on forage grasses (<96%; Table 1). When comparing the *Xtg* core genome to non-*graminis X. translucens* strains, a set of 74 coding sequences (CDS) were found to be unique to *Xtg*. Thirty of these were functionally annotated and were assigned to four categories, i.e. nutrient acquisition, regulation & modification, virulence and adhesion & motility (Hersemann et al. 2017). These unique CDS may play an important role in the plant-pathogen interaction. In addition, when comparing surface exposed structures such as components of the T3SS or the type IV pilus, high sequence deviations were found for *Xtg* when compared to the other *X. translucens* strains. This may allow *Xtg* to evade plant perception and enable successful colonization of *L. multiflorum* (Hersemann et al. 2017).

4   Remaining Challenges

The candidate scaffolds for bacterial wilt resistance identified in *L. multiflorum*, together with the bacterial virulence traits identified in *Xtg* represent with no doubt an invaluable resource for the development of genome assisted selection strategies for bacterial wilt resistance. However, in order to fully understand plant resistance, which seems to be controlled by complex general regulating mechanisms rather than effector triggered immunity responses, the interplay of bacterial virulence traits and plant resistance mechanisms needs to be further investigated. In this context, dual RNA-seq may offer a valuable tool to simultaneously investigate host and pathogen transcriptomes (Westermann et al. 2016). Another potential limitation lies in the phenotypic characterisation of the host-pathogen interaction. So far, all studies relied on scoring of disease symptoms (wilting) on clonally propagated plants. Not only does this method often yield variable data within the same genotype (unpublished data), it most likely also lacks the sensitivity needed to detect subtle changes in the early stages of pathogen infection.

Acknowledgments. This research was partially funded by the Swiss National Science Foundation (grants no. 065417, 112582 and 13858).

References

Byrne SL, Nagy I, Pfeifer M, Armstead I, Swain S, Studer B, Mayer K, Campbell JD, Czaban A, Hentrup S (2015) A synteny-based draft genome sequence of the forage grass *Lolium perenne*. Plant J 84:816–826

Egli T, Goto M, Schmidt D (1975) Bacterial wilt, a new forage grass disease. J Phytopathol 82:111–121

Goodwin S, McPherson JD, McCombie WR (2016) Coming of age: ten years of next-generation sequencing technologies. Nat Rev Genet 17:333–351

Hersemann L, Wibberg D, Blom J, Goesmann A, Widmer F, Vorhölter FJ, Kölliker R (2017) Comparative genomics of host adaptive traits in *Xanthomonas translucens* pv. *graminis*. BMC Genomics 18:35

Knorst V, Byrne S, Asp T, Studer B, Widmer F, Kölliker R (2016) Wisdom of crowds: pooled sequencing identifies genomic regions associated with disease resistance in ryegrass. In: Kölliker R, Boller B (eds) Plant Breeding: The Art of Bringing Science to Life; Abstracts of the 20th Eucarpia General Congress. Agroscope, Zurich, p 10

Kölliker R, Kraehenbuehl R, Boller B, Widmer F (2006) Genetic diversity and pathogenicity of the grass pathogen *Xanthomonas translucens* pv. *graminis*. Syst Appl Microbiol 29:109–119

Michel VV (2001) Interactions between *Xanthomonas campestris* pv. *graminis* strains and meadow fescue and Italian rye grass cultivars. Plant Dis 85:538–542

Pritchard L, Birch PRJ (2014) The zigzag model of plant–microbe interactions: is it time to move on? Mol Plant Pathol 15:865–870

Studer B, Boller B, Herrmann D, Bauer E, Posselt UK, Widmer F, Kölliker R (2006) Genetic mapping reveals a single major QTL for bacterial wilt resistance in Italian ryegrass (*Lolium multiflorum* Lam.). Theor Appl Genet 113:661–671

Suter D, Frick R, Hirschi H-U, Bertossa M (2017) Liste der empfohlenen Sorten von Futterpflanzen 2017–2018. Agrarforschung 8:1–16

Westermann AJ, Förstner KU, Amman F, Barquist L, Chao Y, Schulte LN, Müller L, Reinhardt R, Stadler PF, Vogel J (2016) Dual RNA-seq unveils noncoding RNA functions in host–pathogen interactions. Nature 529:496–501

Wichmann F, Asp T, Widmer F, Kölliker R (2011a) Transcriptional responses of Italian ryegrass during interaction with *Xanthomonas translucens* pv. *graminis* reveal novel candidate genes for bacterial wilt resistance. Theor Appl Genet 122:567–579

Wichmann F, Hug BM, Widmer F, Boller B, Studer B, Kolliker R (2011b) Phenotypic and molecular genetic characterization indicate no major race-specific interactions between *Xanthomonas translucens* pv. *graminis* and *Lolium multiflorum*. Plant Pathol 60:314–324

Wichmann F, Vorhölter F-J, Hersemann L, Widmer F, Blom J, Niehaus K, Reinhard S, Conradin C, Kölliker R (2013) The noncanonical type III secretion system of *Xanthomonas translucens* pv. *graminis* is essential for forage grass infection. Mol Plant Pathol 14:576–588

Genotyping of Festulolium Cultivars Involved in EUCARPIA Multi-site Trial Using DArT Markers and GISH

D. Kopecký[1]([✉]), J. Baert[2], S. Barth[3], J. Bartoš[1], V. Černoch[4],
J. Doležel[1], D. Grogan[5], J. Harper[6], M. Humphreys[6], T. Książczyk[7],
L. Østrem[8], E. Paszkowski[9], D. Sokolović[10], Z. Zwierzykowski[7],
and M. Ghesquière[11]

[1] Institute of Experimental Botany, Centre of the Region Hana
for Biotechnological and Agricultural Research, Šlechtitelů 31,
78371 Olomouc, Czech Republic
kopecky@ueb.cas.cz
[2] ILVO, Caritasstraat 21, 9090 Melle, Belgium
[3] Oak Park Crops Research Centre, Teagasc, Carlow, Ireland
[4] DLF Seeds, s.r.o., Fulnecká 95, 742 47 Hladké Životice, Czech Republic
[5] Department of Agriculture, Food and Marine, Davis Street, Tipperary, Ireland
[6] IBERS, Aberystwyth University, Aberystwyth SY23 3EE, UK
[7] Institute of Plant Genetics, Strzeszynska 34, 60-479 Poznan, Poland
[8] NIBIO, Fureneset, 6967 Hellevik i Fjaler, Norway
[9] DANKO Plant Breeding Ltd., Szelejewo Drugie, 64-820 Piaski, Poland
[10] Institute of Forage Crops, 37251 Globoder, Kruševac, Serbia
[11] INRA, 86600 Lusignan, France

Abstract. A comprehensive set of Festulolium cultivars from on-going field trials in the Eucarpia network was characterised at the chromosome level using genomic *in situ* hybridization (GISH) and by Diversity Array Technology (DArT) markers. Both technologies were found to be complementary in describing the breeding history of the plant material. The genomic composition of the *Lolium* × *Festuca* cultivars varied from those that comprised equivalent proportions of their parental genomes to introgression lines where small chromosome segments of *Festuca* had been translocated onto *Lolium* chromosomes. The breadth of genotype combinations found within the grass cultivars described represents an important resource of genetic variations necessary to combat the diverse abiotic stresses encountered within Europe, including safeguards against prolonged exposure to harsh weather conditions. It is likely that in future plant breeding, genotyping will contribute to precision-transfers of targeted *Festuca* genes into *Lolium* germplasm in order to enhance resilience to climate change.

Keywords: Festulolium · Interspecific hybridization · Genotyping
Genetic variability · Genome composition

© Springer International Publishing AG, part of Springer Nature 2018
G. Brazauskas et al. (Eds.): *Breeding Grasses and Protein Crops in the Era of Genomics*, pp. 155–159, 2018.
https://doi.org/10.1007/978-3-319-89578-9_28

1 Introduction

Successful merging of two genomes in one organism via the process of interspecific hybridization is usually accompanied by polyploidization to encourage homologous chromosome pairing, disomic inheritance and the restoration of complete or at least partial fertility. In order to maintain allopolyploid integrity, the exchange of chromosome segments via homoeologous chromosome pairing and recombination may be hindered by chromosome pairing control systems or by distinctiveness between the parental genomes on the DNA sequence level. This results in the exclusive pairing of homologous chromosomes and thereby genome stability over successive generations (Ramsey and Schemske 1998). However, genomic stability is frequently not evident in Festulolium allopolyploids (i.e. *Festuca* × *Lolium* hybrids). The chromosomes of *Festuca* and *Lolium* pair freely with each other and frequent homoeologous chromosome exchanges are evidenced in successive generations (Kopecký et al. 2005; Zwierzykowski et al. 2006). Studies on genomic composition and gene expression revealed a shift towards the *Lolium* genome, where over generations, chromosomes of *Festuca* are replaced by those of *Lolium* (Zwierzykowski et al. 2006; Kopecký et al. 2017). A similar trend has been identified at the RNA level, where overexpression of *Lolium* alleles over those of *Festuca* was observed for many genes in reciprocal *Festuca* ♀ × *Lolium* and *Lolium* ♀ × *Festuca* hybrids (Stočes et al., in preparation). Thus, no maternal effect has been evidenced on transcriptome level. Moreover, for many important agronomic traits in Festulolium hybrids, *Lolium*-type morphology is much more evident than *Festuca*-type (Humphreys et al. 2014).

The aim of this study under the umbrella of the Eucarpia Fodder Crops and Amenity Grasses Section, Festulolium Working Group was to investigate the phenotypic traits of an elite selection of 15 Festulolium cultivars and breeding material using parental species as controls grown in multi-location field trials across Europe. Over three successive years, the locations contrasted in climate from moderately harsh winter conditions found in Norway to the summer heat and drought found in Serbia. Alongside the field trials, a detailed characterization of the genomic composition of the cultivars was undertaken using genomic *in situ* hybridization (GISH) and a Diversity Arrays Technology (DArT) approach. Through the combined phenotyping/genotyping approach, it was possible to estimate the potential benefits from using genomic composition analysis to predict the field performance of Festulolium hybrids. The genotyping of the individual cultivars is described in this paper.

2 Material and Methods

2.1 Plant Material

The official multi-site field trial consisted of 15 Festulolium cultivars or breeding material namely Lueur (*L. multiflorum* × *F. glaucescens*), Lofa, Becva and FuRs0352 (*L. multiflorum* × *F. arundinacea*), Perseus, Achilles, Perun, Hostyn, AberNiche, Felopa, Sulino and Agula (*L. multiflorum* × *F. pratensis*) and Prior, Fabel and FuRs0142 (*L. perenne* × *F. pratensis*) together with six cultivars of parental species;

L. multiflorum, L. perenne, F. pratensis and *F. arundinacea*. The trials were sown in 2012 in Norway, UK (2 locations), France, Poland, Czech Republic, Serbia, and in 2013 in Belgium and Ireland, and maintained over 3 years.

2.2   Genomic *in situ* Hybridization (GISH)

Genomic *in situ* hybridization was carried out on 20–30 randomly selected plants per cultivar to estimate the level of chromosomal variability within and between cultivars and to determine their genomic composition. The methods used are as described in Kopecký et al. (2005).

2.3   Diversity Arrays Technology (DArT) Genotyping

DArT analysis was undertaken as described in Kopecký et al. (2009). Each cultivar was represented by 20 randomly selected plants, collected as bulked samples. The markers reported herein were selected with a call rate >80% and with technical reproducibility of >99.0%. We estimated the genomic constitution of the Festulolium cultivars and the respective proportions derived from their *Lolium* and *Festuca* parental genomes using DArT markers found to be species-specific in the cultivars of parental species used as controls in the multi-site field trial experiment. The only exception was DArT markers considered to be specific to *F. glaucescens*. This parent was absent in the field trial. However, *F. glaucescens* with *F. pratensis* are the progenitor species of *F. arundinacea*, so DArT markers present in *F. arundinacea* but absent in *F. pratensis* were considered to be derived from *F. glaucescens* genome.

3   Results and Discussion

During the multisite field trial, a number of traits were evaluated over three years including annual and seasonal variation of yield, disease susceptibility and biochemical traits for feeding value. However, the focus here concerns the genome composition of the Festulolium cultivars with the aim to later relate this to their overall field performance across locations.

GISH revealed that all Festulolium cultivars had hybrid chromosome constitution, although not all plants within each cultivar showed the presence of chromatin derived from both *Lolium* and *Festuca* parental species. In cvs. Becva, Lofa and FuRs0352, all developed from *L. multiflorum* × *F. arundinacea* hybrids followed by backcrossing onto tetraploid *L. multiflorum*, we found a large proportion of plants with no evidence for the presence of any *Festuca* chromatin (26 out of 27 and 23 out of 29, respectively). This indicates that the process of backcrossing had led to the rapid elimination of the *Festuca* chromatin and that there may have been preferential transmission of *Lolium* chromosomes and/or possible pre- or postzygotic advantage gained by genotypes having predominance of *Lolium* genome complement. On the other hand, cv. Lueur derived from a *L. multiflorum* × *F. glaucescens* hybrid had almost equal proportions of the parental genomes. The shift in proportion of parental genomes on chromosome level towards *Lolium* was much less than in *L. multiflorum* × *F. pratensis* cultivars.

Thus, we may assume that this cultivar will display greater genome stability in terms of equal proportions of parental genomes likely to be found in successive generations. Cultivars originating from *L. perenne* × *F. pratensis* and *L. multiflorum* × *F. pratensis* hybrids again displayed the predominance of the *Lolium* genome, but still retained considerable amounts of the *Festuca* genome in all the plants analyzed. Homoeologous recombination appeared to be more frequent in *L. multiflorum* × *F. pratensis* than in *L. perenne* × *F. pratensis* hybrids, which indicates a closer relationship between *F. pratensis* and *L. multiflorum* than between *F. pratensis* and *L. perenne*. The proportion of parental chromatin varied from cultivar to cultivar (results shown in Fig. 1).

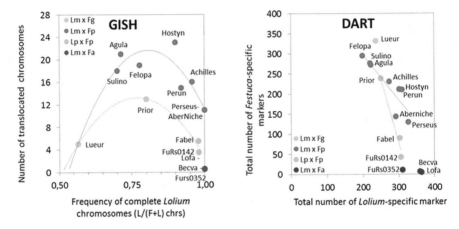

Fig. 1. Genomic composition of tetraploid (2n = 4× = 28) Festulolium cultivars using chromosome (GISH) and DNA level (DArT markers). Lines are quadratic (GISH) and linear (DArT) regression models for *L. multiflorum* × *F. pratensis* and *L. perenne* × *F. pratensis* cultivars.

Using the DArTFest array, we were able to screen all Festulolium cultivars using 821 *Festuca* sp. markers and 412 *Lolium* sp. markers. The results on genomic composition were highly consistent with those from GISH. They confirmed the presence of both parental genomes in all Festulolium cultivars in various proportions (Fig. 1). Whilst GISH chromosome sorting revealed high numbers of translocated chromosomes (with the highest number observed in cv. Hostyn: 24, on average), the negative slope between the number of *Lolium* and *Festuca* markers indicates that *Festuca* loci are being continuously lost to the benefit of the *Lolium* loci. Consistent with the cytological study, *Festuca*-specific markers in *L. perenne* × *F. pratensis* cultivars are lost at a higher rate than in the *L. multiflorum* × *F. pratensis* cultivars. The overall rate of *Festuca* genome introgression among the Festulolium cultivars fitted remarkably well between GISH and DArT, assuming that translocated chromosomes are composed of equal proportions of parental chromatin (r^2 = 0.93, P < 0.002). However, it is interesting to note that cv. Hostyn, although highly similar in genome composition to cvs. Achilles and Perun, has more translocated chromosomes. This suggests that cv. Hostyn retains much of its *Festuca* genome in the form of short introgressions on translocated chromosomes while cvs. Achilles or Perun appear to retain more intact *Lolium* chromosomes. This may indicate greater genome stability in successive generations of cv. Hostyn, an ideal precursor for further selection.

Our preliminary results suggest that the phenotype of Festulolium cultivars in various locations may be predicted to some extent based on the numbers of species-specific markers present and their relative proportions. For example, persistency (estimated as the yield of regrowth after summer in the third year) correlated positively with the introgression rates of the *Festuca* genome in locations subject to stresses where the fescue controls performed better than those of ryegrass (Festulolium Working Group, in preparation). The results also indicate that an amphiploid breeding approach with a high *Festuca* genome content is necessary in locations where the stresses are particularly harsh. Interestingly, markers of both genome sources were also found to be positively associated with persistency in more advanced introgressed cultivars which emphasizes the potential of genome selection within Festulolium for improving resilience of grasslands (Festulolium Working Group, in preparation).

4   Conclusion

To conclude, our study revealed large variation in genome composition of Festulolium cultivars. Using GISH and DArT marker technologies, we could monitor proportions of parental chromatin in hybrids and observed the predominance of *Lolium* genome over that of *Festuca* in all cultivars. Only in *L. multiflorum* × *F. glaucescens* cv. Lueur was there near-equivalent content of parental chromatin. Genotyping provides predictive potential for certain phenotypic traits, and might speed-up the breeding process, and assist the evaluation of existing cultivars.

Acknowledgements. DK, JB and JD were supported by the National Program of Sustainability I (award LO1204).

References

Humphreys MW, O'Donovan SA, Farrell MS, Gay A, Kingston-Smith AL (2014) The potential of novel Festulolium (2n = 4× = 28) hybrids as productive, nutrient-use-efficient fodder for ruminants. J Food Energy Secur, 1–13. https://doi.org/10.1002/fes3.50

Kopecký D, Lukaszewski AJ, Doležel J (2005) Genomic constitution of Festulolium cultivars released in the Czech Republic. Plant Breed 124:454–458

Kopecký D, Bartoš J, Lukaszewski AJ, et al. (2009) Development and mapping of DArT markers within the Festuca-Lolium complex. BMC Genom 10: 473

Kopecký D, Šimoníková D, Ghesquière M, Doležel J (2017) Stability of genome composition and recombination between homoeologous chromosomes in Festulolium (*Festuca* × *Lolium*) cultivars. Cytogenet Genome Res 151:106–114

Ramsey J, Schemske DW (1998) Pathways, mechanisms and rates of polyploid formation in flowering plants. Annu Rev Ecol Syst 29:467–501

Zwierzykowski Z, Kosmala A, Zwierzykowska E, Jones N, Joks W, Bocianowski J (2006) Genome balance in six successive generations of the allotetraploid *Festuca pratensis* × *Lolium perenne*. Theor Appl Genet 113:539–547

Transcriptomic Analysis of Drought-Sensitive and Tolerant Genotypes of Perennial Ryegrass (*Lolium perenne* L.)

M. O. Leyva[1], I. Nagy[1], S. Hentrup[1], S. Byrne[2], and T. Asp[1(✉)]

[1] Molecular Biology and Genetics, Aarhus University, Aarhus, Denmark
torben.asp@mbg.au.dk
[2] Crop Science Department, Teagasc, Oak Park, Carlow, Ireland

Abstract. Ten perennial ryegrass genotypes were tested for response to water deficiency in greenhouse under controlled conditions. Four genotypes with contrasting response to drought were selected for RNAseq transcriptomic analysis. In total, we found 5321 differentially expressed genes (DEGs) during drought but only 232 DEGs were shared across all genotypes. We could identify 7 genes with contrasting expression profiles in two tolerant genotypes compared to the sensitive genotypes. Most of these genes code for regulatory proteins that can be utilized for future selection strategies in breeding material for enhanced drought tolerance.

Keywords: *Lolium perenne* · Drought · RNAseq

1 Introduction

Climate change models predict greater variability in rainfall patterns and increased periods of summer drought will affect grasslands. Therefore, it is important to develop new forage crop cultivars that can adapt to such climatic changes. Stress tolerance mechanisms are aimed at protecting against cellular damage when avoidance mechanisms are no longer sufficient. Plant drought tolerance is a complex quantitative trait, involving multiple metabolic pathways (Claeys and Inzé 2013). These mechanisms include detoxification of reactive oxygen species and the accumulation of protective proteins like late embryogenesis abundant (LEA) proteins and solutes such as proline. Avoidance and tolerance responses are mainly orchestrated by the hormone abscisic acid (ABA) and drought responsive element binding (DREB) proteins. On the other hand, transcription factors such as gibberellin signalling DELLA proteins and ethylene response factors, the reprogramming of proline and a mitochondrial metabolism are nodes in the co-regulation of growth and drought tolerance. These nodes are promising targets for breeding of drought-tolerant plants. Most of the molecular knowledge about drought resistance has been acquired for model plants. However, there are less genetic data available for perennial ryegrass. Candidate gene association mapping of drought tolerance identified a putative LpLEA3, an iron superoxide dismutase (LpFeSOD) and a copper-zinc superoxide dismutase (Yu et al. 2013). Liu and Jiang (2010) found 256 expressed sequence tags (ESTs) using a suppression subtractive hybridization

© Springer International Publishing AG, part of Springer Nature 2018
G. Brazauskas et al. (Eds.): *Breeding Grasses and Protein Crops in the Era of Genomics*, pp. 160–165, 2018.
https://doi.org/10.1007/978-3-319-89578-9_29

approach. About 10% of the 95 candidate genes with known functions are of detoxification functions. Recently, a heat shock protein, a FeSOD and an aquaporin (PIP1) have been found to be downregulated during low watering (Jiang et al. 2016). In this work, Illumina RNA sequencing have been used to compare the response of tolerant and sensitive genotypes to get robust transcriptomic data to identify common and specific elements of the molecular mechanism of drought stress.

2   Material and Methods

Vegetatively propagated shoots of ten pre-selected genotypes were grown under controlled conditions in a greenhouse (20 °C, 16 h light). The relative humidity in soil (RH) was recorded in each individual pot using chip-based sensors (Flowerpower, Parrot). Four pots per genotype were kept as controls and plants in eight pots were adapted to 50% of the control plants RH. After this, watering was stopped for drought-treated plants. Total RNA was extracted according to manufacturer instructions (Sigma, STRN250) from leaf samples in three biological replicates from drought-treated and control plants at 0 h, 48 h, 96 h and 144 h. RNAseq data was obtained for 95 samples. The reads were mapped onto reference sequences from a comprehensive transcript database of perennial ryegrass (Byrne et al. 2015) using the Trinity software pipeline. The different time points were compared to all the control samples. A minimum fold change $\equiv$ FC $\geq$ 4 and p-value cutoff for FDR $\leq$ 0.0001 was chosen to filter differentially expressed genes (DEGs). InterProscan was used to identify conserved structural elements and to assign Gene Ontology information to predicted protein sequences.

3   Results and Discussion

3.1   Determination of Drought Tolerance

No drought symptoms were observed before watering was stopped (Fig. 1). After that point, the phenotypic characterization allowed the identification of two tolerant and two sensitive genotypes.

Tolerant 1 and Sensitive 1 displayed the most tolerant and the most sensitive phenotype.

3.2   Transcriptomic Response to Drought

Comparing the drought samples vs. control samples for every genotype, we identified a total of 5321 drought-DEGs but only 232 DEGs were common for all genotypes (Fig. 2A).

These 232 DEGs were all up-regulated during drought. The Gene Ontology analysis of these genes compared with the genes expressed in control plants revealed that six biological processes were overrepresented in the common response to drought (Fig. 2B). The most significant were "embryo development" represented by five LEA

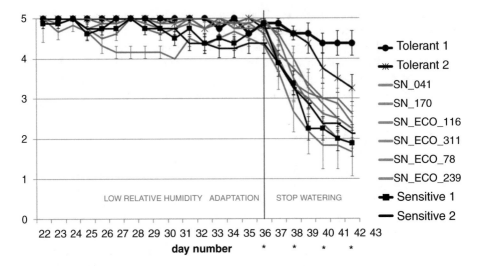

Fig. 1. Average phenotypic scores (n = 8) during the drought experiment for ten perennial ryegrass genotypes where *5* refers to a turgid plant, *4* – not turgid leaves, *3* – leaves starting to curl, *2* – first dry leaves, *1*– most part of the plant dry, *0* – no visible green. (*) sampling days.

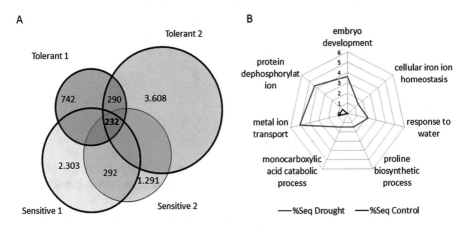

Fig. 2. A. Venn diagram representing the number of drought DEGs shared by the four genotypes analyzed. B. GO terms overrepresented in the common drought-DEGs compared to the genes expressed in control plants (p < 0.05).

proteins group 1 type, "protein dephosphorylation" represented by six PPM-type serine/threonine phosphatase 2C similar to *abi1* and *abi2* ABA receptors and "cellular iron ion homeostasis" represented by ferritins and transmembrane proteins containing cation-binding and transport domains. We also found 2 genes involved in proline synthesis and 3 putative dehydrins involved in "response to water" which code for LEA group 2 type proteins.

3.3 Genes with Contrasting Expression Level Between Tolerant and Sensitive Lines

Genes with similar expression profile in both tolerant genotypes but different from sensitive genotypes were filtered. As a result, seven DEGs were identified (Fig. 3). Three DEGs had higher expression level in tolerant lines compared to the sensitives (Fig. 3A–C). In Arabidopsis, FHY3 (FAR-RED ELONGATED HYPOCOTYL3) and

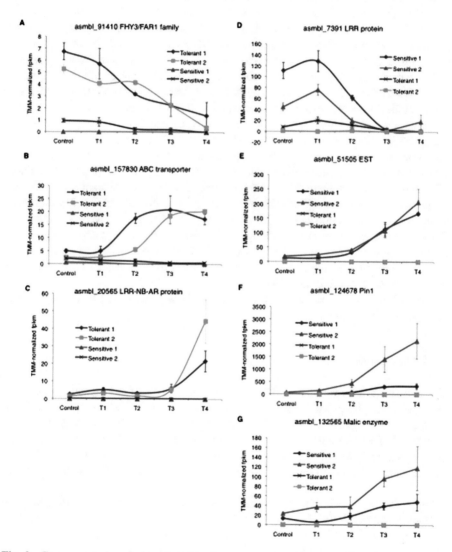

Fig. 3. Gene expression during drought of seven genes with contrasting expression level between tolerant and sensitive genotypes. *Control* – optimally watered plants, *T1* – half relative humidity adapted plants, *T2* – 48 h after stopping watering, *T3* – 96 h after stopping watering, *T4* – 120 h after stopping watering.

FAR1 (FAR-RED IMPAIRED RESPONSE1) are positive regulators of ABA signaling (Tang et al. 2013). In perennial ryegrass we found this transcript to be downregulated during drought but it was highly expressed in tolerant control plants compared to sensitive plants (Fig. 3A). This result supports that this gene could be a key negative regulator of downstream drought response in tolerant genotypes. On the other hand, a Rli/ABCE1 transporter gene was induced at early drought stages (Fig. 3B). In yeast, Rli1/ABCE1 controls translation re-initiation in 3'UTRs *in vivo* and ribosome homeostasis. In maize it has been found that the translational efficiencies of several genes changed significantly in response to drought (Lei et al. 2015). The results support this gene as a key translation factor for drought resistance in perennial ryegrass. Another putative transcription factor belonging to the NB-ARC class of Leucine Rich Repeat (LRR) proteins was found to be over-expressed at the final stages of drought treatment in the experiment (Fig. 3C). In Arabidopsis, constitutive or conditional enhanced expression of the CC–NBS–LRR gene ADR1 conferred significant drought tolerance (Chini et al. 2004). Another LRR-gene was found but had higher expression level in sensitive genotypes compared to tolerant genotypes and was downregulated during drought (Fig. 3D). A putative function of this protein could be as a repressor of other regulatory proteins that activate the response to drought. Other 3 genes had higher expression level in sensitive lines compared to the tolerant but were drought-induced (Fig. 3E–G). One of them is an uncharacterized expressed sequence of perennial ryegrass (Accession nr: EY458053, Fig. 3E). Two others are coding for a protease inhibitor (*Pin1*) and malic enzyme, respectively (Fig. 3F, and G). Over-expression of a *Pin* gene resulted in improved drought resistance at the reproductive stage under drought field conditions in rice (Huang et al. 2007). The expression of malic enzyme from maize resulted in altered stomatal behaviour and reduced water loss in transgenic tobacco plants (Laporte et al. 2002).

According to the results in the present work, these genes might be drought related stress factors in perennial ryegrass.

4   Conclusion

The transcriptomic response to drought has a high level of intraspecific variability. LEA proteins, phosphatases 2C similar to *abi1* and *abi2* ABA receptors, ferritins, proline synthesis genes and dehydrins respond to drought in all genotypes. The malic enzyme, *Pin1* and a transcript of unknown function are putative drought stress factors only expressed in sensitive genotypes. We hypothesize that FHY3/FAR1, Rli/ABCE1 and LRR-NB-ARC could be key regulators in drought tolerance in perennial ryegrass.

Acknowledgments. This work was supported by the project for "Improving crop yields and production properties" (FAUPE), granted by Promilleafgiftsfonden.

References

Byrne SL, Nagy I, Pfeifer M, Armstead I, Swain S, Studer B, Mayer K, Campbell JD, Czaban A, Hentrup S, Panitz F, Bendixen C, Hedegaard J, Caccamo M, Asp T (2015) A synteny-based draft genome sequence of the forage grass *Lolium perenne*. Plant J 84:816–826

Chini A, Grant JJ, Seki M, Shinozaki K, Loake GJ (2004) Drought tolerance established by enhanced expression of the CC–NBS–LRR gene, ADR1, requires salicylic acid, EDS1 and ABI1. Plant J 38(5):810–822

Claeys H, Inzé D (2013) The agony of choice: how plants balance growth and survival under water-limiting conditions. Plant Physiol 162(4):1768–1779

Huang Y, Xiao B, Xiong L (2007) Characterization of a stress responsive proteinase inhibitor gene with positive effect in improving drought resistance in rice. Planta 226(1):73–85

Jiang Y, Cui Y, Pei Z, Liu H, Sun S (2016) Growth response and gene expression to deficit irrigation and recovery of two perennial ryegrass accessions contrasting in drought tolerance. HortScience 51(7):921–926

Laporte MM, Shen B, Tarczynski MC (2002) Engineering for drought avoidance: expression of maize NADP-malic enzyme in tobacco results in altered stomatal function. J Exp Bot 53(369):699–705

Lei L, Shi J, Chen J, Zhang M, Sun S, Xie S, Li X, Zeng B, Peng L, Hauck A (2015) Ribosome profiling reveals dynamic translational landscape in maize seedlings under drought stress. Plant J 84(6):1206–1218

Liu S, Jiang Y (2010) Identification of differentially expressed genes under drought stress in perennial ryegrass. Physiol Plant 139(4):375–387

Tang W, Ji Q, Huang Y, Jiang Z, Bao M, Wang H, Lin R (2013) FAR-RED ELONGATED HYPOCOTYL3 and FAR-RED IMPAIRED RESPONSE1 transcription factors integrate light and abscisic acid signaling in Arabidopsis. Plant Physiol 163(2):857–866

Yu X, Bai G, Liu S, Luo N, Wang Y, Richmond DS, Pijut PM, Jackson SA, Yu J, Jiang Y (2013) Association of candidate genes with drought tolerance traits in diverse perennial ryegrass accessions. J Exp Bot 64(6):1537–1551

Selection of Perennial Ryegrass Spaced Plants in Sward Conditions

J. Aper[✉], A. Ghesquiere, and J. Baert

Plant Sciences Unit, Flanders Research Institute for Agriculture,
Fisheries and Food, Melle, Belgium
jonas.aper@ilvo.vlaanderen.be

Abstract. We evaluated growth, winter damage and rust susceptibility of diploid and tetraploid perennial ryegrass genotypes that were cloned and planted as spaced plants in three 'environments': on bare soil, in a red fescue sward and in a white clover sward. Scores were most repeatable on bare soil. Over all, correlation was higher between the scores from plants on bare soil and in red fescue than between the scores on bare soil and in white clover. Some accessions showed different growth reactions in one of the three swards.

Keywords: *Lolium perenne* · Competition · Persistence

1 Introduction

Progress in breeding perennial ryegrass (*Lolium perenne* L.) is slow and this has been attributed to the low correlation between the growth of spaced plants in the first breeding phases and the final yield performance of the progeny in a dense sward (Conaghan and Calser 2011). Evaluating spaced plants in a sward might improve this correlation. Van Dijk and Winkelhorst (1978) evaluated spaced perennial ryegrass plants in swards of different grass species and concluded that an evaluation in sward conditions eliminates all weak plants and the percentage of surviving plants gave a good indication of persistence. Spaced plants in sward conditions are more easy to be kept weed-free and neighbour effects of other spaced plants are strongly decreased. The aim of this study was to evaluate the growth of diploid and tetraploid perennial ryegrass genotypes that were cloned and planted in three conditions: bare soil, a red fescue sward and a white clover sward.

2 Materials and Methods

2.1 Timeline

On 22 November 2013, 10 populations of diploid perennial ryegrass and 10 populations of tetraploid perennial ryegrass were sown in trays in the greenhouse (Table 1). For each population, one tray with 96 seeds was sown. On 24 January 2014, 50 plants with the highest number of tillers were selected and planted in pots of $7 \times 7 \times 8$ cm. Between 24 and 28 March 2014, the plants with the highest number of tillers were again selected

© Springer International Publishing AG, part of Springer Nature 2018
G. Brazauskas et al. (Eds.): *Breeding Grasses and Protein Crops in the Era of Genomics*, pp. 166–170, 2018.
https://doi.org/10.1007/978-3-319-89578-9_30

and 30 plants of each population were cloned in 2–4 ramets in a new tray. On 14 April, the plants were again cloned to end up with 6 ramets per plant in trays. In total there were 10 populations × 30 plants per population × 6 ramets per plant = 1800 plants for each ploidy level (diploid – tetraploid). Between 20 and 23 May 2014, the plants were planted in 2 blocks in a split plot design with sward as main plot factor and genotype as subplot factor. The plants were planted in 3 swards: bare soil, red fescue and white clover. Diploid and tetraploid plants formed 2 separate subblocks within one block-sward combination. The 300 plants for each ploidy level were randomized within one subblock. The plants were spaced 50 cm in the row × 40 cm between rows. Each 4 rows were flanked on each side by a row of the crown rust-susceptible perennial ryegrass variety 'Melvina'. The swards were installed by sowing the red fescue variety 'Rolf' at a turf grass sowing density of 10 kg per 300 m^2 and the white clover variety 'Merwi' at 0.15 kg per 300 m^2. During several observation moments in 2014, 2015 and 2016, the plants were visually scored using an ordinary scale from 1 (bad) to 5 (good) (Table 2).

Table 1. Number of genotypes of the 10 populations of diploid and tetraploid perennial ryegrass used in the evaluation trial

Diploid populations				Tetraploid populations			
Name	Origin	HD	Number of genotypes	Name	Origin	HD	Number of genotypes
Abermagic	Variety	Inter	30	6138	Population	Early	26
Arsenal	Variety	Inter	28	6169/3	HS family	Early	28
6144	Population	Inter	28	6169/9	HS family	Early	29
6170/1	HS family	Inter	28	Maurizio	Variety	Inter	26
Aberavon	Variety	Late	29	6185	Population	Inter	27
6183	Population	Late	24	6175/5	HS family	Inter	30
6180/3	HS family	Late	26	6177/3	HS family	Inter	28
6180/4	HS family	Late	30	6177/9	HS family	Inter	15
6182/3	HS family	Late	28	Aberbite	Variety	Late	20
6182/5	HS family	Late	30	Polim	Variety	Late	25

HD - heading date, *HS* - half sib

Table 2. Observation moments[a]

	2014		2015		2016	
Observation	D	T	D	T	D	T
Crown rust resistance[b]	13/10	17/10				
Absence of winter damage			23/03	23/03	18/03	21/03
Spring growth			20/04	13/04	22/04	28/04
Summer/autumn growth	26/10	17/10	26/10	23/10	12/08	22/08

[a] *D*: diploid, *T*: tetraploid
[b] Only scores for plants in red fescue and bare soil

2.2   Fertilisation/Mowing Regime

The experiment was mown three times in 2014 and 2016 and five times in 2015. The bare soil plots and these with red fescue were fertilized with 218 kg N, 51 kg P_2O_5 and 275 kg K_2O/ha. The plots covered with white clover received a fertilization of 68 kg N, 72 kg P_2O_5 and 240 kg K_2O/ha.

2.3   Statistical Analysis

The diploid and tetraploid populations are analysed separately. After planting, 20 diploid plants out of 1800 (1.1%) and 76 tetraploid plants out of 1800 (4.2%) did not establish. All the plants with less than 6 ramets were not included in the analysis. In total 281 diploid and 254 tetraploid genotypes remained for the statistical analysis (Table 1).

The absolute difference of the scores between the 2 ramets of a plant for each of the 3 swards was calculated. The effect of the sward was evaluated with a Wilcoxon rank-sum test with a Bonferroni correction when more than 2 averages were compared. Next, we calculated Spearman rank correlations between the different swards with a selected set of observations (averaged over the 2 ramets of a plant). The average final growth score in August 2016 was evaluated for each population with a Wilcoxon rank-sum test with a Bonferroni correction.

3   Results and Discussion

Table 3 shows the absolute difference between 2 ramets of the same plant, averaged over the 281 diploid and 254 tetraploid genotypes in the same sward. For crown rust resistance, bare soil and red fescue were similar in reproducibility for diploid plants, but bare soil was more precise than red fescue for tetraploid plants. The general order in reproducibility was bare soil > red fescue > white clover. The differences were larger in 2016 when compared with 2015.

The correlations between the different observations and the three swards are shown in Table 4. Over all, the correlation was higher between the scores on plants on bare soil and in red fescue, compared with the correlation between the scores on plants on bare soil and in white clover. The highest correlations were found between autumn growth 2014 and spring growth 2015 in the three swards.

Table 5 shows the summer growth in August 2016 (=2 years and 3 months after planting). As expected, the average growth score for a population was lower in red fescue and white clover because of competition. But some populations reacted differently in the two swards: the growth of Arsenal was not significantly reduced in red fescue, while 6180/3 had high average growth scores in the 3 swards. For the tetraploids, the average growth score of Polim, 6175/5 and 6177/3 was significantly reduced in white clover, compared to red fescue. This was confirmed on genotype level when a hierarchical ANOVA analysis was performed and a significant interaction between clone and sward was found (results not shown).

Table 3. Average absolute difference between visual scores on 2 ramets of the same plant[a,b,c]

		D			T		
Observation	Year	BS	RF	WC	BS	RF	WC
Crown rust resistance[d]	2014	0.51a	0.40a	–	0.59a	0.88b	–
Absence of winter damage	2015	0.44a	0.90b	0.81b	0.46a	0.66ab	0.82b
	2016	0.70a	0.99b	1.22b	0.78a	0.77a	1.27b
Spring growth	2015	0.60a	0.78a	0.81a	0.70a	0.80ab	0.98b
	2016	0.80a	1.11ab	1.19b	0.68a	0.84a	1.21b
Autumn growth	2014	0.73a	0.77a	1.17b	0.63a	0.72a	0.70a
	2015	0.65a	0.94b	1.05b	0.78ab	0.66a	0.98b
	2016	0.70a	1.34b	1.59b	0.78a	0.97a	1.06a

[a] D – diploid, T – tetraploid, BS – bare soil, RF – red fescue, WC – white clover
[b] Values followed by the same letter within the same observation and within the same ploidy level are not significantly different according to a Wilcoxon Rank Sum test ($p < 0.01$ with Bonferroni correction when 3 averages were compared).
[c] Visual scores on a scale of 1 to 5: 1 = bad, 5 = good
[d] Only scores for plants in red fescue and bare soil

Table 4. Spearman rank correlation coefficients between selected scores in different swards (above diagonal: diploid, below diagonal: tetraploid)[a,b]

		BS				RF				WC		
		AG 2014	CR 2014	SG 2015	AG 2016	AG 2014	CR 2014	SG 2015	AG 2016	AG 2014	SG 2015	AG 2016
BS	AG 2014		0.43	0.63	0.41	0.58	0.31	0.55	0.43	0.30	0.45	0.35
	CR 2014	0.46		0.37	0.38	0.45	0.58	0.39	0.33	0.22	0.29	0.26
	SG 2015	0.72	0.31		0.52	0.5	0.29	0.61	0.46	0.28	0.54	0.43
	AG 2016	0.45	0.31	0.52		0.42	0.33	0.38	0.47	0.22	0.36	0.45
RF	AG 2014	0.48	0.21	0.51	0.39		0.45	0.69	0.55	0.45	0.52	0.36
	CR 2014	0.28	0.53	0.29	0.23	0.29		0.39	0.33	0.27	0.29	0.21
	SG 2015	0.55	0.26	0.67	0.46	0.62	0.34		0.63	0.37	0.59	0.37
	AG 2016	0.52	0.30	0.54	0.63	0.56	0.29	0.61		0.31	0.48	0.37
WC	AG 2014	0.36	*0.14*	0.41	0.32	0.47	0.23	0.4	0.38		0.70	0.24
	SG 2015	0.54	0.25	0.67	0.52	0.52	0.31	0.67	0.58	0.65		0.43
	AG 2016	0.39	0.22	0.41	0.54	0.32	0.26	0.38	0.48	0.43	0.57	

[a] BS – bare soil, RF – red fescue, WC – white clover, AG – autumn growth, CR – crown rust resistance, SG – spring growth
[b] Values in italic are significant at $p < 0.05$, all other values are significant at $p < 0.01$.

Between May and June 2016, plants were scored for heading date and leaf colour and we made 12 polycrosses with selections coming from swards of white clover and red fescue and from bare soil. The half-sib families coming from these polycrosses will be evaluated in a yield trial to compare the different selection 'environments'.

Table 5. Summer growth (August 2016) of 20 diploid and tetraploid perennial ryegrass populations, averaged over the number of plants per population[a,b]

Diploid populations				Tetraploid populations			
Name	BS	RF	WC	Name	BS	RF	WC
Abermagic	3.38a	1.42b	1.82b	Maurizio	2.17a	0.83b	0.44b
Arsenal	2.61a	1.77ab	1.68b	Aberbite	1.70a	0.38b	0.43b
6144	3.18a	1.55b	1.82b	6138	2.81a	1.33b	0.81b
6170/1	3.46a	1.86b	2.21b	6185	3.35a	1.69b	1.09b
Aberavon	3.05a	1.38b	1.36b	6169/3	3.86a	2.16b	1.68b
6183	2.88a	1.67b	1.52b	Polim	2.64a	1.20b	0.42c
6180/3	3.88a	3.00b	3.23ab	6169/9	3.90a	1.98b	1.47b
6180/4	3.72a	2.73b	2.32b	6175/5	3.40a	2.28b	1.08c
6182/3	3.75a	2.63b	2.66b	6177/3	3.55a	2.21b	1.11c
6182/5	4.05a	2.75b	2.72b	6177/9	2.77a	0.77b	0.40b

[a] D – diploid, T – tetraploid, BS – bare soil, RF – red fescue, WC – white clover

[b] Values followed by the same letter within the same population and within the same ploidy level are not significantly different according to a Wilcoxon Rank Sum test ($p < 0.01$ with Bonferroni correction)

4 Conclusion

The highest reproducible results were found for scores on plants on bare soil. The scores on plants in red fescue had a higher correlation with the scores on plants on bare soil than with the scores on plants in white clover. Growth of several accessions was more reduced in the white clover sward when compared with the red fescue sward.

Acknowledgments. We are grateful for Geert Lejeune for the heading date scores and the excellent assistance of the forage breeding group for careful field maintenance.

References

Conaghan P, Calser MD (2011) A theoretical and practical analysis of the optimum breeding system for perennial ryegrass. Ir J Agric Food Res 50:47–63
Van Dijk GE, Winkelhorst GD (1978) Testing perennial ryegrass (*Lolium perenne* L.) as spaced plants in swards. Euphytica 27:855–860

Variation in Intrinsic Water Use Efficiency Between Perennial Ryegrass Genotypes Differing for Drought Tolerance

P. Westermeier[1]([✉]), R. Schäufele[2], and S. Hartmann[1]

[1] Institute for Crop Science and Plant Breeding, Bavarian State Research Center
for Agriculture, Freising, Germany
peter.westermeier@lfl.bayern.de
[2] Grassland Group, Technische Universität München, Freising, Germany

Abstract. Drought tolerance of perennial ryegrass (*Lolium perenne* L.) is one of the constraints for the use of this forage species in drought prone areas. Global climate change with increasing fluctuation in rainfall patterns will further limit its use in larger areas of Europe. The aim of this study was to investigate whether there is genotypic variation for carbon isotope discrimination in drought tolerance of selected perennial ryegrass as a measure for water use efficiency. The phenotypic selection for drought tolerance is not generally correlated with carbon isotope composition within the material investigated. For one genotype with good phenotypic drought tolerance a significant difference in intrinsic water use efficiency was found under stress conditions. It is also shown, that water use efficiency is not the only drought tolerance mechanism within perennial ryegrass, as other drought tolerant genotypes show no significant differences in intrinsic water use efficiency but a good recovery after drought stress at the same time.

Keywords: *Lolium perenne* · Carbon isotope discrimination
Intrinsic water use efficiency · Genetic variability

1 Introduction

Perennial ryegrass (*Lolium perenne* L.; PR) is extensively used for forage use due to high nutritional quality in livestock feeding and high productivity. On the other hand PR has no distinct drought tolerance, thus global climate change will limit its use in some areas due to increased drought periods and shifts in rainfall patterns. Selection for drought tolerance in crop plants is often hampered by suitable selection environments and a low heritability of the target trait "yield under drought stress". Thus, secondary selection traits are needed with higher heritability which enable a higher selection gain. One of the widely used selection traits for yield under drought stress is water use efficiency (WUE). Measuring this trait is error prone and laborious, thus assessing the discrimination against ^{13}C ($\Delta^{13}C$) is the mean of choice, as $\Delta^{13}C$ is negatively correlated with WUE (Farquhar and Richards 1984). In bread wheat (*Triticum aestivum* L.) $\Delta^{13}C$ has been used successfully for breeding drought tolerant varieties (Rebetzke et al. 2002). Genotypic variation for $\delta^{13}C$ was also found in different forage grass species such as PR

© Springer International Publishing AG, part of Springer Nature 2018
G. Brazauskas et al. (Eds.): *Breeding Grasses and Protein Crops in the Era of Genomics*, pp. 171–175, 2018.
https://doi.org/10.1007/978-3-319-89578-9_31

and orchard grass (*Dactylis glomerata* L.) (Jensen et al. 2002) as well as in crested wheatgrass (*Agropyron desertorum* (Fischer ex Link) Schultes) and tall fescue (*Festuca arundinacea* Schreb.) (Johnson and Bassett 1991).

2   Material and Methods

2.1   Plant Material and Phenotyping

Ten PR clones originating from breeding material and genebank accessions were investigated in a controlled greenhouse environment as a pot experiment. Diploid PR clones (tolerant: 36_21; 36_37; 81_13; 82_1; 170_30; 170_39; susceptible: 108_18; 112_38; 132_4; 132_13) were preselected for divergent drought stress response in a two-year rain-out shelter experiment according to visual biomass scorings. One representative of *Festulolium* (FEL) and meadow fescue (MF; *Festuca pratensis* L.) each, showing drought tolerance in the field, were added to the set for comparison. The greenhouse trial was performed as a randomized complete block design, with three replicates and two treatments (well-watered control vs. drought stress). Plants were cultivated in standard potting substrate (75% black peat, 25% white peat, 9 vol. % clay granulate, 8 kg/m^3 bentonite) in 15 × 15 cm pots. At 2015-07-07 (cutting timepoint 3) all pots were filled to a maximum water content, for the next 44 days (cutting timepoint 4) TDR measured soil water content in the stress treatment was reduced from 64.8% to approx. 3%, controlled by repeated measurements and water supply to keep stress level comparable between all pots. Watering was completely withhold until day 72 (cutting timepoint 5) which resulted in a volumetric water content of nearly 0%. Above ground biomass was determined after 72 days. For determining carbon isotope composition (δ ^{13}C) biomass samples were oven dried (60 °C), grinded, and an aliquot of 0.7 mg was analyzed in a Delta Plus, Thermo Finnigan MAT (Bremen, Deutschland) mass spectrometer, coupled to an elementar analyzer (NA1110, Carlo Erba Instruments). Each sample was measured against a working gas standard, calibrated against a secondary isotope standard. Δ ^{13}C (‰) was calculated according to Farquhar et al. (1982) as ($\delta_a - \delta_p$)/(1000 + δ_p)*1000 with δ_a as the δ ^{13}C (‰) value of the ambient atmosphere (-10‰ constant) and δ_p as the δ ^{13}C (‰) values of the plant samples. Discrimination of plants with C$_3$ photosynthesis can be modeled by the following equation Δ ^{13}C = a + (b − a)*(c$_i$/c$_a$) (Farquhar et al. 1982). As factors a (discrimination against ^{13}CO$_2$ due to slower diffusion = 4.4‰) and b (discrimination against ^{13}CO$_2$ by RuBisCO enzyme = 28‰) are constant there is a linear correlation between Δ^{13}C and c$_i$/c$_a$ (concentration of CO$_2$ in stomata c$_i$; concentration of CO$_2$ in the atmosphere c$_a$ = 400 ppm) and c$_i$ can be calculated from known Δ^{13}C. WUE$_i$ (the ratio of net assimilation to stomatal conductance) is calculated as c$_a$/(1-c$_i$/c$_a$)/1.6 (Farquhar and Richards 1984). The software package Plabstat version 3A (Utz 2011) was used for ANOVA analysis, calculation of repeatability, adjusted means and least significant difference (LSD). Genotype 132_13 was excluded from statistical analysis due to abnormal growth behavior during the whole experiment.

3 Results and Discussion

Significant genotypic variation for the traits biomass production and WUE_i in PR was found under well-watered as well as under drought stress conditions. Repeatability for the traits were between 68.2% and 81.2% under well-watered conditions and between 75.5% and 86.6% under drought stress (Table 1). Biomass production dramatically declined under stress conditions to a mean of 13% compared to well-watered control, repeatability of WUE_i under drought stress increased compared to control. Correlation for dry matter biomass yield between well-watered and stressed plants was 45.9%, for WUE_i 69.5% (Fig. 1). No significant correlation was detectable between WUE_i and biomass dry matter neither under well-watered nor und stress conditions.

Table 1. Variation for yield (biomass DM) and intrinsic water use efficiency (WUE_i; μmol CO_2 mol^{-1} H_2O) within selected perennial ryegrass genotypes.

Treatment	Trait	Mean	Min	Max	LSD 5%	Rep [%]
Control	Biomass DM [g]	2.39	0.23	5.16	1.21	81.2 **
	WUE_i	40.19	27.68	55.16	7.61	68.2 *
Stress	Biomass DM [g]	0.31	0.05	0.86	0.22	75.5 **
	WUE_i	73.81	57.23	89.69	7.89	86.6 **

* sign. genotypic variance at $P < 0.05$; ** sign. genotypic variance at $P < 0.01$

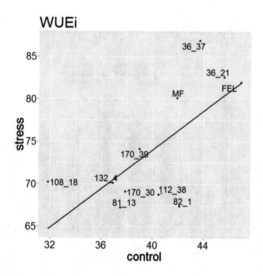

Fig. 1. Correlation of WUE_i (μmol CO_2 mol^{-1} H_2O) under well-watered and drought conditions $r = 69.5\%$

174   P. Westermeier et al.

Figure 2 shows the WUE$_i$ of investigated genotypes under well-watered and drought conditions. Especially genotypes 36_21 and 36_37 showed, beside the drought tolerant species FEL and MF a lower $\Delta^{13}C$, indicating higher WUE$_i$. This is in accordance with the theory of carbon isotope discrimination. For plants with C$_3$ photosynthetic pathway, $\Delta^{13}C$ is a well suited tool to assess long-term WUE, as it is assimilation weighted mean integrating over the complete growth period (Condon et al. 2004). Potentially drought tolerant plants close their stomata at an earlier stage of drought stress to prevent additional water losses. As net assimilation is not reduced to the same extent in these plants, c_i/c_a decreases, subsequently decreasing $\Delta^{13}C$ (Farquhar et al. 1989).

In other crop species like wheat a quantitative inheritance of carbon isotope discrimination was found with small genotype-by-environment interactions and high heritability (Condon and Richards 1992). Rebetzke et al. (2002) found also significant correlations between reduced carbon isotope discrimination and an increase in grain yield of bread wheat under rain-fed conditions in Australia. This work resulted in two Australian wheat cultivars (Rees and Drysdale) with enhanced water use efficiency detected by carbon isotope discrimination (Condon et al. 2002). Similar results were reported by Johnson and Bassett (1991) for four different cool-season grasses (including perennial ryegrass) with consistent results for carbon isotope discrimination across environments. Jensen et al. (2002) found a higher variability for carbon isotope discrimination in PR than in orchard grass (*Dactylis glomerata* L.). For PR carbon isotope discrimination was found to be higher in diploid cultivars than in tetraploids across five levels of water supply indicating, that tetraploid material is potentially more drought tolerant than diploids.

There was no general correlation between the preselection for drought tolerance in PR and the carbon isotope composition/water use efficiency. In a previous study recovery after drought was identified as the most important trait for drought tolerance

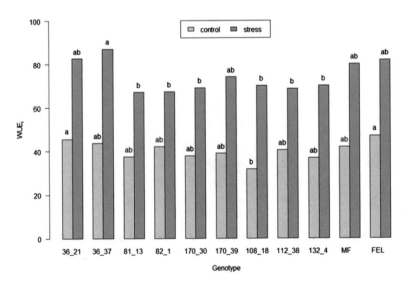

Fig. 2. Intrinsic water use (μmol CO$_2$ mol^{-1} H$_2$O) efficiency of single genotypes under contrasting water supply.

within the selected material (Westermeier and Hartmann 2016). But this trait enables a combination of good recovery after temporary drought stress and optimized water use efficiency for breeding new PR varieties with a good performance under temporary drought stress conditions.

4 Conclusion

Beside regrowth rate after drought water use efficiency seems to play an important role in drought tolerance of perennial ryegrass. Significant genotypic variation for WUE_i could be found within the presented set of perennial ryegrass clones. As the correlation of WUE_i betweeen well-watered and stress conditions is good, it seems to be feasible to select for enhanced WUE_i under normal growth conditions to draw conclusions on drought tolerance. Results need to be proved on a larger set of plants. For this purpose segregating crossing populations were built based on the plant material shown here which is phenotyped for drought tolerance and WUE_i.

Acknowledgments. This study was funded by the Bavarian State Ministry of Food, Agriculture and Forestry (grant E/15/01).

References

Condon AG, Richards RA (1992) Broad sense heritability and genotype x environment interaction for carbon isotope discrimination in field-grown wheat. Aust J Agric Res 43:921–934

Condon AG, Richards RA, Rebetzke GJ, Farquhar GD (2002) Improving intrinsic water-use efficiency and crop yield. Crop Sci 42:122–131

Condon AG, Richards RA, Rebetzke GJ, Farquhar GD (2004) Breeding for high water-use efficiency. J Exp Bot 55:2447–2460

Farquhar GD, Ehleringer JR, Hubick KT (1989) Carbon isotope discrimination and photosynthesis. Annu Rev Plant Physiol Plant Mol Biol 40:503–537

Farquhar GD, O'Leary MH, Berry JA (1982) On the relationship between carbon isotope discrimination and the intercellular carbon dioxide concentrations in leaves. Aust J Plant Physiol 9:121–137

Farquhar GD, Richards RA (1984) Isotopic composition of plant carbon correlates with water-use efficiency of wheat genotypes. Aust J Plant Physiol 11:539–552

Jensen KB, Asay KH, Johnson DA, Waldron BL (2002) Carbon isotope discrimination in orchardgrass and ryegrasses at four irrigation levels. Crop Sci 42:1498–1503

Johnson RC, Basset LM (1991) Carbon isotope discrimination and water use efficiency in four cool-season grasses. Crop Sci 31:157–162

Rebetzke GJ, Condon AG, Richards RA, Farquhar GD (2002) Selection for reduced carbon isotope discrimination increases aerial biomass and grain yield of rainfed bread wheat. Crop Sci 42:739–745

Utz HF (2011) PLABSTAT - A computer program for statistical analysis of plant breeding experiments. Version 3A. Germany, Universität Hohenheim

Westermeier P, Hartmann S (2016) Varying growth behavior of *Lolium perenne* L. clones under drought conditions and after rewatering. Grassl Sci Eur 21:832–834

Physiological Indicators of Tolerance to Soil Water Deficit in *Lolium multiflorum/Festuca arundinacea* Introgression Forms

K. Masajada[✉], A. Augustyniak, D. Perlikowski, D. Ratajczak,
W. Zwierzykowski, I. Pawłowicz, and A. Kosmala

Institute of Plant Genetics, Polish Academy of Sciences, Poznan, Poland
kmas@igr.poznan.pl

Abstract. *Festuca arundinacea* is a model plant species for a group of forage grasses, comprising the *Lolium-Festuca* complex to recognize the crucial mechanisms of tolerance to water deficit and recovery after stress cessation. The *L. multiflorum/F. arundinacea* introgression forms are excellent plant materials for dissecting these important traits into several components. Here, four genotypes of the introgression forms with a different capacity in response to abiotic and biotic stresses, were analyzed with respect to their physiological performance under water deficit conditions and further recovery. It was revealed that the Relative Water Content and Electrolyte Leakage parameters could be good indicators of tolerance to soil water deficit and ability to recover after stress cessation in the *L. multiflorum/F. arundinacea* introgression forms.

Keywords: Electrolyte leakage · *Festuca arundinacea* · *Lolium multiflorum*
Relative water content · Tolerance to water deficit

1 Introduction

Lolium multiflorum (Italian ryegrass) is forage grass species characterized by a high forage quality, but a relatively low tolerance to biotic and abiotic environmental stresses. On the other hand, *Festuca arundinacea* (tall fescue) is species with a high level of abiotic stress tolerance, particularly tolerance to water deficit (Kosmala et al. 2012), but it does not match *L. multiflorum* regarding productivity and quality in short term leys. The hybridization of both species enables the assembly of their complementary characters within a single genotype. The *L. multiflorum/F. arundinacea* introgression forms were shown earlier to be excellent plant models for dissecting drought tolerance of *F. arundinacea* into several crucial components (Perlikowski et al. 2014, 2016a, 2016b). Here, four closely related BC_5 introgression forms of *L. multiflorum/F. arundinacea* with significant differences in their frost tolerance, winter-hardiness and resistance to *Microdochium nivale* (Augustyniak et al. 2015) were used to evaluate their tolerance to soil water deficit. To do this, several physiological parameters were measured during simulated water deficit conditions followed by a recovery period.

© Springer International Publishing AG, part of Springer Nature 2018
G. Brazauskas et al. (Eds.): *Breeding Grasses and Protein Crops in the Era of Genomics*, pp. 176–180, 2018.
https://doi.org/10.1007/978-3-319-89578-9_32

2   Material and Methods

2.1   Plant Material and Stress Treatment

Plant materials used in the present research involved four *L. multiflorum/F. arundinacea* introgression forms obtained after five rounds of backcrossing of *L. multiflorum* (4x) × *F. arundinacea* (6x) hybrid to *L. multiflorum* (4x) (Perlikowski et al. 2014). These plants were selected earlier from a larger population with respect to their winter hardiness in natural field conditions as well as frost tolerance and resistance to *Microdochium nivale* in simulated laboratory conditions (Table 1) (Augustyniak et al. 2015, 2016). These four forms, each one in four biological replicates, were transferred to pots (1.75 dm^3), containing a sand:peat (1:3) mixture. The experiment of 11 days of water deficit (no water was applied to pots), followed by 10 days of re-watering was performed in a growth chamber at a temperature of 22 °C (16 h day/8 h night, light of 400 μmol(quanta) m^{-2} s^{-1}, HPS "Agro" lamps, Philips, Brussels, Belgium) and 30% relative air humidity.

Table 1. Characteristics of *L. multiflorum/F. arundinacea* introgression forms

Genotype	Winter hardiness	Frost tolerance	Resistance to *Microdochium nivale*
180/30/19	++	−	++
180/30/75	++	+	++
180/30/84	−	+	−
180/30/138	−	++	−

Trait intensity: (++) high, (+) low, (−) lack of stress tolerance/resistance

2.2   Physiological Parameter Measurements

The physiological parameters, including relative water content (RWC), electrolyte leakage (EL), and chlorophyll fluorescence were analyzed before stress treatment (control), at three different time-points of progressing water deficit (after three, six and 11 days), and after 10 days of subsequent re-watering.

The RWC in the leaves was measured to monitor the leaf hydration status. This parameter was calculated as follows: RWC% = (FW−DW)/(SW−DW) × 100, where FW is the leaf fresh weight, DW is the leaf dry weight, and SW is the leaf turgid weight. Leaf FW was weighed in a leaf segment of 3 cm in length cut from the middle part of the leaves immediately after leaf collection from plants. Next, the leaves were submerged overnight in water-containing closed test tubes, dried with filter paper and weighed (SW). To measure DW leaves were dried at 70 °C for one day.

The EL was calculated according to the following formula: L_1/L_2 × 100. Leaf segments (two of 2 cm in length) cut from the middle parts of the leaves were placed in closed test tubes containing 15 cm^3 of distilled deionized, sterile water. They were incubated overnight at room temperature, while gently shaken. Electrical conductivity (L_1) was measured using a conductivity meter (Hanna Instruments EC215 Conductivity Meter). Leaves were then frozen in liquid nitrogen, placed again in the same test tubes, and L_2 was measured after overnight shaking.

Chlorophyll 'a' fluorescence was measured in the youngest, but fully expanded leaves, by means of the HandyPEA fluorimeter (Hansatech Instruments Ltd., King's Lynn, England) during midday as described in detail by Kosmala et al. (2012) and Perlikowski et al. (2014).

The analyses of the physiological parameters were performed in four biological replicates of each genotype (10 leaves per one replicate). Means for replicates and standard errors (SE) of RWC and EL are shown in the Fig. 1 and 2, respectively. A two-way analysis of variance (ANOVA), with genotype and time-point as classification factors, has been performed. Differences in RWC, EL and chlorophyll fluorescence parameters between the plants during experiment duration were evaluated using Tukey HSD test at P = 0.05.

3   Results and Discussion

The significant drop of RWC was observed for all the analyzed genotypes on the 11th day of stress treatment. However, some differences among the introgression forms also existed. The highest alterations in water content were observed for the genotype 180/30/84, and the lowest–for 180/30/138. This fact indicated that the latter genotype was less disturbed by a water deficit stress. On the other hand, after further re-watering, no differences with respect to the RWC were observed among the genotypes, indicating also a relatively high level of water status regeneration in the genotype 180/30/84 (Fig. 1).

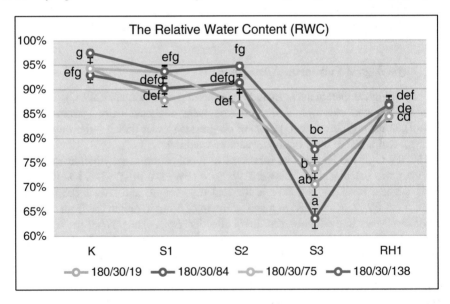

Fig. 1. The Relative Water Content (RWC) in four *L. multiflorum/F. arundinacea* introgression forms before stress treatment (K), on the 3rd (S1), 6th (S2), and 11th (S3) day of water deficit and 10 days after re-hydration (RH) initiation. The data represent means for ten individual leaves per one biological replicate (40 leaves in total per genotype) ± SE. Values of the same parameters marked with the same letter did not differ significantly at P = 0.05 according to Tukey HSD test.

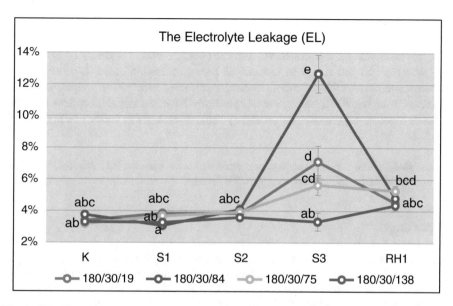

Fig. 2. The Electrolyte Leakage (EL) in four *L. multiflorum/F. arundinacea* introgression forms before stress treatment (K), on the 3rd (S1), 6th (S2), and 11th (S3) day of water deficit and 10 days after re-hydration (RH) initiation. The data represent means for ten individual leaves per one biological replicate (40 leaves in total per genotype) ± SE. Values of the same parameters marked with the same letter did not differ significantly at P = 0.05 according to Tukey HSD test.

The tendency observed for the RWC was also noticed for the EL parameter. Its value was not changed during the whole experiment in the genotype 180/30/138 but increased significantly on the 11th day of water deficit to the highest degree in the genotype 180/30/84 (Fig. 2).

However, no significant differences between the genotype 180/30/138 and 180/30/84 with respect to the level of EL after re-hydration were observed, indicating a high capacity of the latter introgression form to regenerate damaged cellular membranes (Fig. 2). On the other hand, the genotype 180/30/138 revealed a high level of membrane stability under the stress condition (Fig. 2), but the leaf tissue dehydration was also lowest under stress conditions in this genotype (Fig. 1). Interestingly, the genotype 180/30/138 was shown to be the most frost tolerant plant among the four analyzed in our earlier study (Augustyniak et al. 2015).

The majority of all the analyzed chlorophyll fluorescence parameters during the whole experiments did not reveal any significant differences among the four introgression forms, indicating the similar performance of photosystem II in the analyzed plants during water deficit and recovery period.

4 Conclusion

The physiological parameters–RWC and EL could be good indicators of tolerance to soil water deficit and ability to recover after stress cessation in the *L. multiflorum/F. arundinacea* introgression forms as it was demonstrated earlier for *F. arundinacea*. The plants with higher RWC and lower EL generally have leaf metabolism less disturbed by stress conditions, thus it means that they better tolerate water deficit in soil (Kosmala et al. 2012).

Acknowledgments. The research was performed within a project funded by the Polish Ministry of Agriculture and Rural Development (no. 35; 2016).

References

Augustyniak A, Perlikowski D, Pociecha E, Dziurka M, Płażek A, Rapacz M, Kosmala A (2015) Performance of lolium multiflorum/festuca arundinacea introgression forms under abiotic and biotic stress conditions. Abstracts of the 10th international conference of plant functioning under environmental stress, 16–19 September, 2015, Cracow, Poland, p. 39

Augustyniak A, Masajada K, Płażek A, Pociecha E, Dziurka M, Kosmala A (2016) Molecular indicators of resistance to Microdochium nivale in Lolium multiflorum/Festuca arundinacea introgression forms. Abstracts of the 20th eucarpia general congress of eucarpia, 29 August–1 September, 2016, Zurich, Schwitzerland, p. 246

Kosmala A, Perlikowski D, Pawłowicz I, Rapacz M (2012) Changes in the chloroplast proteome following water deficit and subsequent watering in a high and a low drought tolerant genotype of Festuca arundinacea. J Exp Bot 63:6161–6172

Perlikowski D, Kosmala A, Rapacz M, Kościelniak J, Pawłowicz I, Zwierzykowski Z (2014) Influence of short-term drought conditions and subsequent re-watering on the physiology and proteome of Lolium multiflorum/Festuca arundinacea introgression forms with contrasting levels of tolerance to long-term drought. Plant Biol 16:385–394

Perlikowski D, Czyżniejewski M, Marczak Ł, Augustyniak A, Kosmala A (2016a) Water deficit affects primary metabolism differently in two Lolium multiflorum/Festuca arundinacea introgression forms with a distinct capacity for photosynthesis and membrane regeneration. Front Plant Sci 7:1063. https://doi.org/10.3389/fpls.2016.01063

Perlikowski D, Kierszniowska S, Sawikowska A, Krajewski P, Rapacz M, Eckhardt Ä, Kosmala A (2016b) Remodeling of leaf cellular glycerolipid composition under drought and re-hydration conditions in grasses from the *Lolium-Festuca* complex. Front Plant Sci 7:1027. https://doi.org/10.3389/fpls.2016.01027

Incidence of Six Grass Species by *Fusarium* sp. as a Cause of Silvertop

O. Trněný[(⊠)] and J. Nedělník

Agricultural Research, Ltd., Troubsko, Czech Republic
trneny@vupt.cz

Abstract. Silvertop is the big threat to grass seed production because of its devastating effect on yield. Screening of three *Fusarium* species in silvertop stalks confirmed interdependence between symptoms of diseases and *Fusarium poae* infection. Pooled samples of six grass species were tested for the presence of infection by PCR detection of *F. poae*, *F. graminearum* and *F. culmorum*. *F. poae* were detected in 67% of silvertop pooled samples and only in 8% in asymptomatic pooled samples. *F. culmorum* were not detected in any case and *F. graminearum* were not detected at a significant level in comparisons silvertop and asymptomatic samples.

Keywords: Silvertop · Grass disease · *Fusarium poae*
Fusarium graminearum · *Fusarium culmorum* · PCR detection

1 Introduction

Silvertop, a disease affecting fertile stems and inflorescences of certain cultured and wild grass species (also known as white head or white ear) results in loss of the ability to create developed seeds. It remains an important disease in grass seed production and can cause even total seed production losses. The disease's main symptom is complete drying and necrosis of the fertile plant stem at an early development stage. After grasses head, inflorescences cease to develop and do not blossom. The fertile stems remain short, dry, and become yellowish or silvery in color, markedly differing from the previously green stems and later straw-coloured mature stems. The infected stem's overall appearance is described as similar to a shepherd's crook due to its characteristic curvature (Cagaš 2004). As the most frequent biotic causal agents of so-called parasitic silvertop, the literature mentions various insect species as well as mites as vectors (Mühle et al. 1971) acting together with the fungus *Fusarium poae*. This type of silvertop is characterized by shrinking, darkening, and subsequent necrosis of the infected stem in the basal section of the last internode. The disease primarily affects older stands and is certainly one of the most serious diseases for grasses grown for seed (Cagaš 2004).

The main aim of this study was to observe the level of infection of *Fusarium* fungi by PCR methods in peduncles from six grass species as potential crude factors of disease plants.

© Springer International Publishing AG, part of Springer Nature 2018
G. Brazauskas et al. (Eds.): *Breeding Grasses and Protein Crops in the Era of Genomics*, pp. 181–185, 2018.
https://doi.org/10.1007/978-3-319-89578-9_33

2 Material and Methods

2.1 Plant Material and PCR Detection

In total 318 grass stalks (150 symptomatic and 168 asymptomatic) of six grass species (*Poa compresa* L., *Poa nemoralis* L., *Poa palusrtis* L., *Trisetum flavescens* (L.) P. Beauv., *Lolium perenne* L. and *Cynosurus cristatus* L.) were used for PCR diagnostic of *F. poae*, *F. graminearum* and *F. culmorum* infection. All samples were obtained from the production grass seeds parcels from OSEVA PRO LtD. (Zubří, Czech Republic, 49.4652839N, 18.0781656E). The objects of diagnosis were approximately 2 cm pieces of peduncles from highest nods above. Because of avoiding outside contamination peduncles were pulled out from adjacent leaves. Preparation of samples was easy for silvertop plants because the connection was disrupted between peduncles and the highest nods but it was quite cumbersome for asymptomatic plants. Samples were pooled to 54 bulks. From these bulks, 30 were formed by pooling of 5 pieces of silvertop stalks and 24 bulks were formed by pooling of 7 pieces of asymptomatic stalks from different plants of one species. Plant material was homogenized in liquid nitrogen using a mortar and pestle and total DNA was isolated from each bulk by DNeasy Plant Mini Kit (Qiagen, Germany) according to the standard kit protocol. A quality of DNA isolations and results of PCR amplification were analysed using 1% agarose gel with EtBr.

Diagnostic PCRs contain 7.5 µl 2x PPP Master mix (Top-Bio, Czech Republic), 0.4 µM each primer, 1 µl DNA and top up with PCR water to 15 µl. PCR conditions and used primers are listed in Table 1. Because of validation of results three types of controls were used. Negative control, positive control with DNA from the clear isolate

Table 1. Used conditions and primers sequences for diagnostic PCR

Fusarium species	Primers	Reference	PCR conditions
Fusarium poae	5'ACCGAATCTCAACTCCGCTTT'3 5'GTCTGTCAAGCATGTTAGCACAAGT'3	Nicolaisen et al. 2009	94 °C 5 min; **35x** (94 °C 30 s; 62 °C 30 s; 72 °C 40 s); 72 °C 10 min
Fusarium graminearum	5'CCATTCCCTGGGCGCT'3 5'CCTATTGACAGGTGGTTAGTGACTGG'3	Nicolaisen et al. 2009	94 °C 5 min; **35x** (94 °C 30 s; 61 °C 45 s; 72 °C 40 s); 72 °C 10 min
Fusarium culmorum	5'GATGCCAGACCAAGACGAAG'3 5'GATGCCAGACGCACTAAGAT'3	Schilling et al. 1996	94 °C 5 min; **35x** (94 °C 30 s; 55 °C 40 s; 72 °C 45 s); 72 °C 10 min

of all three *Fusarium* species from CCM Czech Collection of Microorganisms (Masaryk University, Brno, Czech Republic) and positive amplification control for each samples using *Poaceae*-specific primers (Poa-gen-S541 5'GCTTTCTCATTCTACT-CTTTC'3, Poa-gen-A551 5'CTTTTCTTGTGCATCATCCTAG'3) and the following PCR conditions: 94 °C 5 min; 35 × (94 °C 30 s; 56 °C 45 s; 72 °C 40 s); 72 °C 5 min (Wallinger et al. 2012).

In order to statistical analysis of the diagnostic PCR screening results, Fisher's exact test for count data was used. P-value and odds ratio were computed using function "fisher.test" in R software (R Core Team 2015).

3 Results and Discussion

Statistically significant difference between symptomatic and asymptomatic samples was obtained only in case of detection *F. poae* (P-value = 0.0001019). Otherwise in both cases of detection *F. culmorum* and *F. graminearum* statistically significant differences were not observed (P value < 0.05). *F. poae* was detected by using PCR in 67% bulks of silvertop samples and in 8% samples of asymptomatic plants. And odds that silvertop samples will be infected by *F. poae* is almost seventeens times higher than odds that asymptomatic plants will be infected by *F. poae* (odds ratio = 16.85).

F. poae infection detected in 100% of silvertop samples and in 0% of asymptomatic samples were only in *P. palustris* L. samples. The grass species that was infected by *F. poae* considerably less in spite of silvertop symptom was *L. perenne* L. (20% silvertop samples were infected by *F. poae*). On the other hand, infection by *F. poae* was detected in asymptomatic samples only in *P. compresa* L. samples (50% asymptomatic samples were infected by *F. poae*) Fig. 1.

F. culmorum were not detected in any samples. *F. graminearum* were detected in only 5 cases of all. Two of these detections were from silvertop samples and three were from asymptomatic samples. And detection occurred in three species (*P. compresa* L. 3 detections, *T. flavescens* (L.)P. Beauv. 1 detection and *L. perenne* L 1 detection).

There are generally several old hypotheses. First of them says that the fungus *F. poae* is a causal agent of parasitic silvertop (Stewart and Hodgkiss, 1908). Another hypothesis previously confirmed in field experiments states that the main vector of *F. poae* transfer to plants is the meadow plant bug (*Leptopterna dolabrata* L.), an insect from the order Hemiptera, family Miridae (Rotrekl et al. 1985). Recent literature sources do not provide sufficient relevant information related to the level of fungal infection in silvertop grasses. Because of the seriousness of the silvertop problem in grass seed production new research is highly desirable to deal with this problem and therefore our results provide a new deeper insight into the relationship between silvertop disease and *Fusarium* infection.

PCR diagnostic screening of association between three *Fusarium* species and silvertop symptoms confirm the hypothesis that silvertop in a grass is associated with *F. poae* infection. But from the results we can conclude that *F. poae* is not only one causal agent of silvertop because one-third of all symptomatic samples was not infected by *F. poae*. On the other hand, neither *F. culmorum* nor *F. graminearum* was diagnosed to a significant level in symptomatic samples. Therefore eventual causal agents of silvertop in one-third of symptomatic samples remains unclear.

The probable cause of extensive disease attack in seed production parcels are insect vectors of pathogenic fungi. Because of this fact insecticides are applied during stem extension phase of growing grasses that are used for seed production (Cagaš and Macháč 2001). Recently Fridrich et al. (2011) found that the intensity of mulching the incidence of silvertop and also the occurrence of nymphs of *L. dolabrata* decreases.

The occurrence of silvertop disease in the Czech Republic was mention in publication Nedělník et al. (2015). In 2013, silvertop was observed at the Zubří location for the first time on a loloid-type intergeneric hybrid of Italian ryegrass (*Lolium multiflorum* Lamk.) and tall fescue (*Festuca arundinacea* Schreber.) (Bečva variety). Microscopic analysis confirmed the presence of *F. poae*. In 2014, this disease was again observed on grasses from the genus *Lolium – L. multiflorum* Lamk. and perennial ryegrass (*L. perenne* L.).

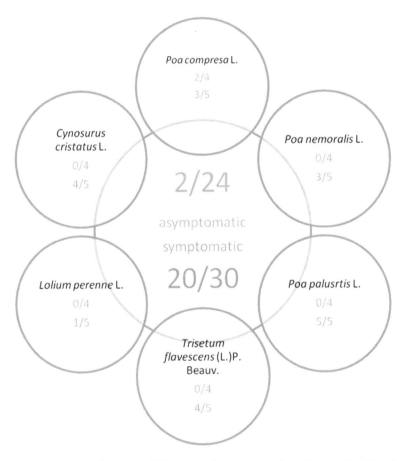

Fig. 1. *Fusarium poae* diagnostic PCR results in symptomatic and asymptomatic silvertop grass samples. Numerators represent a number of positively detected bulks in *F. poae* and denominators represent a number of tested bulks.

4   Conclusion

PCR diagnostic screening of association between three *Fusarium* species and silvertop symptoms confirm the hypothesis that silvertop in a grass is often associated with *F. poae* infection, although there might be other species as *F. poae* was not identified in all silvertop cases. These are unlikely to be *F. culmorum* nor *F. graminearum* but other unidentified species. Also, the vector is important. Therefore, resistance to *F. poae* is important in breeding grass against silvertop but attention is needed also in the vector.

Acknowledgments. This study was supported by institutional funding on the long-term conceptual development of Agricultural Research, Ltd. organisation.

References

Cagaš B, Macháč J (2001) Protection of grasses against weeds, diseases and pests. ÚZPI, 47 s., ISBN: 80-7271-076-1 (in Czech)

Cagaš B (2004) Silvertop – the main disease of grasses for seed production. Úroda. ProfiPress, s. r.o. 7: 40–41 (in Czech)

Frydrych J, Lošák M, Cagaš B, Rotrekl J, Kolařík P, Barták M (2011) The occurrence of parasitic silvertop at the seed grass stands. Rostlinolékař, ProfiPress, s.r.o. 5: 12–15 (in Czech)

Mühle E, Frauenstein K, Schumann K, Wetzel T (1971) Krankheiten und Schädlinge der Futtergräser. Leipzig: S. Hirzel Verlag 422 s

Nedělník J, Strejčková M, Sabolová T, Cagaš B, Both Z, Palicová J, Hortová B (2015) First Report of Fusarium poae Associated with and/or Causing Silvertop on Loloid-type Festulolium in the Czech Republic. Plant Prot Sci 51(3):136–140

Nicolaisen M, Suproniene S, Nielsen LK, Lazzaro I, Spliid NH, Justesen AF (2009) Real-time PCR for quantification of eleven individual Fusarium species in cereals. J Microbiol Methods 76:234–240

R Core Team (2015). R: A language and environment for statistical computing. R Foundation for Statistical Computing, Vienna, Austria. https://www.R-project.org/

Rotrekl J, Klumpar J, Cagaš B, Bumerl J (1985) True bugs and total silvertop of grass. Ochrana Rostlin, ÚVTIZ 21(4):267–274 (in Czech)

Schilling AG, Moler EM, Geiger HH (1996) Polymerase chain reaction-based assays for species-specific detection of Fusarium culmorum. F. graminearum and F. avenaceum. Phytopathol 86:515–522

Wallinger C, Juen A, Staudacher K, Schallhart N, Mitterrutzner E, Steiner EM, Thalinger B, Traugott M (2012) Rapid Plant Identification Using Species- and Group-Specific Primers Targeting Chloroplast DNA. PLoS ONE 7(1):e29473

Crown Rust and Brown Blight Infection of Perennial Ryegrass Cultivars Representing Three Maturity Groups

R. Aavola$^{(\boxtimes)}$, P. Sooväli, S. Tamm, and A. Bender

Estonian Crop Research Institute, Jõgeva, Estonia
rene.aavola@etki.ee

Abstract. Infection of perennial ryegrass with fungal pathogens was assessed during three years on a basis of a variety type. The scoring served a goal of searching for donor varieties for resistance breeding from amongst the most resistant variety types. Six types representing three maturity classes and two ploidy levels were compared. From amongst 33 cultivars, lates Barnhem (2x), Elgon (4x) and Herbie (2x) turned out to be most resistant against crown rust (1.8–1.9 pts). Brown blight proved to be the prevailing disease, which infected the least foliage (3.0–3.5 pts) in cv. Arvicola, Maurice (both 4x) and Aberdart (2x). Eight tetraploid cultivars preceded the most resistant diploid Herbie on the basis of combined resistance against the two fungi. Three healthiest tetraploids were late-maturing. Among the most resistant diploids four were late-maturing and one was intermediate. Direct effects of cultivars and years to the resistance against both diseases were highly significant (p < 0.001) in both ryegrass ploidy groups. Cultivar × year co-effects did not impact (p > 0.05) the brown blight resistance in tetraploid cultivars.

Keywords: *Lolium perenne* · Disease resistance · Diploid · Tetraploid

1 Introduction

Perennial ryegrass (hereafter PRG) has been considered a minor forage grass in Estonia due to its uncertain persistency. It has a number of superior agronomic traits that encourages the farmers to include it to the seed mixtures. Therefore the proportion of PRG in the seed mixtures has gradually increased. This has been enabled by efficient selection towards acceptable winter survival of new cultivars. In addition, the winters have and will become milder in Northern Europe (Kjellström et al. 2011).

Due to many assets PRG is included into the imported and Estonian domestic forage seed mixtures designed for various grassland types. It is seeded into the pasture and silage mixtures with more persistent grass species, clovers and alfalfa. Among the grasses cultivated in Estonia, PRG is characterized by fastest establishment and regrowth, excellent feeding value and palatability, ease of ensilability and highest seed production. Besides uncertain overwintering, spread of fungal diseases from late summer onwards can be another defect of PRG. Severe infections reduce the productivity and palatability of grass. As genotype × environment interaction strongly impacts

© Springer International Publishing AG, part of Springer Nature 2018
G. Brazauskas et al. (Eds.): *Breeding Grasses and Protein Crops in the Era of Genomics*, pp. 186–190, 2018.
https://doi.org/10.1007/978-3-319-89578-9_34

cultivar performance in PRG (Grogan and Gilliland 2011) careful choice of parents with verified disease resistance for crossings is inevitable. The main pathogens that yearly distress PRG in Estonia are *Puccinia coronata* (hereafter *Pc*) and *Drechslera siccans* (hereafter *Ds*). *Pc* is highly specialised to the varieties it will colonize and large variation is noticed among PRG (Pedersen 1979). Rust attacks begin in July, culminate in August–September and decline in October. Long distance dispersal of uredospores is regarded as most common way of infection. Successful establishment is dependent on the presence of atmospheric humidity (Pedersen 1979, Hagan 2005). Heavy attacks reduce herbage yield and quality as well as root development. During heavy attacks the leaves rapidly turn pale yellow and wither completely after few days. According to Lancashire and Latch (1966) the disease can deplete the reserve substances in PRG that leads to a dry matter yield reduction up to 30%. In Estonia, health problems and liver injuries have been registered in dairy calves as a consequence of feeding on PRG severely infected with rust.

Brown blight, caused by *Ds* can completely decompose the mesophyll, leaving only the veins (Pedersen 1979). Diseased plants show many leaves with dead tips, on which the fungus readily sporulates.

Estonian grassland managers could pay more attention to discrimination between variety types of PRG. If foreign varieties express their genetically inherited advantageous traits in local natural conditions, more benefits from PRG can be expected. In view of changing climate, new winter and disease resistant breeding populations of PRG need to be created and adapted. Current field experiment focused on searching for healthy donor varieties for PRG breeding towards improved disease resistance.

2 Material and Methods

A three-year field experiment was carried out at Estonian Crop Research Institute in Jõgeva (58°27' N, 26°46' E, 70 m a.s.l.). The trial was sown in a randomised complete block design onto Calcaric Cambisol in May. Plot size was 6.4 m^2, replicated four times. We tested 16 diploid and 17 tetraploid cultivars named on Fig. 3. Their respective seeding rates were 30 and 45 kg ha^{-1}. Each type of cultivar–early, intermediate and late, either diploid or tetraploid, was represented by 5 or 6 entries.

Before sowing N50-P11-K21 kg ha^{-1} in the form of compound fertiliser Kemira Power 18-9-9 was applied. Each of the repeated top-dressings with the same fertiliser was equal to 60-13-25 kg of NPK ha^{-1}. Fertiliser was applied in spring after the onset of grass growth, further 3–4 days after harvests, except the last harvest in October. The annual rates of NPK totalled 120-26-50 kg ha^{-1} in 2004, 240-52-100 in 2005 and 180-39-75 in 2006.

After appearance, the foliar diseases were scored direct before each harvest of PRG as indicated in Table 1. The visual estimation was based on the percentage of foliage area covered by pustules or lesions of a pathogen. The scores between 1 and 9 mean: 1 = no disease, 2 = trace of disease, 3 = 5%, 4 = 10%, 5 = 25%, 6 = 40%, 7 = 60%, 8 = 75% and 9 = > 75% of the foliage covered with disease.

Significance of the differences between the mean scores was tested at a 95% probability and the contribution of variation sources was computed by ANOVA using

Table 1. Dates of disease infection observations in perennial ryegrass preceding five harvests in three growing seasons

Year	Harvests				
	I	II	III	IV	V
1	–	–	–	31 Aug	6 Oct
2	15 June	27 July	–	1 Sept	14 Oct
3	–	14 July	14 Aug	28 Sept	–

statistical package Agrobase 20. The standard deviations of the means were computed by Microsoft Excel.

3   Results and Discussion

In late August of year 1 the late diploids (LD) were least infected with *Pc* (Fig. 1). Irrespective of few days without rains during next five weeks, the infection declined to nearly a half of the previous scores, whereby intermediate and late tetraploids (IT and LT, respectively) became least susceptible. In year 2, first symptoms of rust were registered in 6 cultivars in late July, gaining 3 pts in Abercraigs and Jaspis (both LT). Thereafter, the severest proliferation of *Pc* followed in 5 weeks, especially in early and intermediate diploids (ED and ID, respectively). Both attained a peak (4.5 pts) during the entire growing season, after a warm, excessively wet period with high air humidity. Although the mean air temperature did not fall below 10 °C in sunny and dry September and intensive growth of PRG lasted during the next 1.5 months, the infection was not as severe in all variety types in October as it was during previous scoring. Most of the late cultivars remained healthy likely due to dry conditions unsuitable for *Pc* proliferation. In year 3 the withered PRG was scored after going through a severe heat and drought stress. After a continuous hot period in the first half of July (10 days with > 27 °C) with very dry air the pustules of *Pc* appeared on only 5

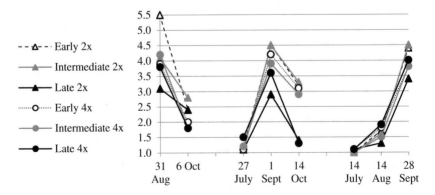

Fig. 1. Crown rust infection of six perennial ryegrass cultivar types in three years. Scale: 1 = no infection, 9 = > 75% covered with symptoms

cultivars in mid-July, most severely in Abercraigs again. Next interim of four weeks was characterised by warm (> 15 °C) nights and the dry spell ended with drought. Therefore 12 cultivars (among them four LDs) still remained healthy in mid-August. As an after-effect of rainy weather in late August, dry but abnormally warm September predisposed the rapid development of *Pc*, whereby LDs were still least and IDs and EDs the most diseased.

Significant differences in resistance against *Pc* in PRG were registered between the two ploidy types in year 1 only, when tetraploids exceeded the resistance of diploids (3.4 pts) by 0.5 pts. In next two years and as an average of the testing cycle, the differences were 0.1 pts at maximum and thus non-significant (p > 0.05). Hence the PRGs at both ploidy levels overall withstood equally the rust infection.

Alike with *Pc*, the moderate infection with brown blight (*Ds*) in year 1 in PRG scaled down during September and first week of October irrespective of abundant precipitation, most in ITs–by 1.6 pts (Fig. 2). Tetraploids withstood *Ds* better (2.6–2.8 pts) at the end of growing season in comparison with diploids (3.2–3.4 pts). Unlike with *Pc*, the first lesions of *Ds* in year 2 appeared on foliage in cool mid-June already. The disease gained a peak (5.5 pts) in year 2 among LDs as early as in late July following the stunted top growth caused by water-deficient soil. The remaining variety types remained less infected even at a time of next two scorings in warm but dry autumn regardless of slight disease progression in EDs, ITs and LTs. In mid-October, as well as in comparison between the yearly means, the tetraploid PRGs maintained significantly better resistance to *Ds*. Year 3 proved that *Ds* can be notably more contagious to PRG than *Pc*. *Ds* could infect PRG even during severe drought, very low air humidity and daily air temperatures exceeding 27 °C. By mid-July as well as by next observations LDs were most susceptible. Opposed to *Pc*, *Ds* made a full use of the warm nights with the air temperatures over 15 °C in August, although the heat had withdrawn and the drought recurred. Compared with the previous year, the disease steadily progressed and peaked evidently in the end of growing season. The score of 6 pts assigned to LDs in September was an ultimate value among both diseases averaged per variety types throughout the entire study. Less than a half of the normal precipitation amount in September did not hold back the outbreaks of both registered diseases, which were fostered by high mean air temperature exceeding the long-term average by 3 °C. The PRGs representing both ploidy types did not differ significantly (p > 0.05) in *Ds* infection in year 3. However,

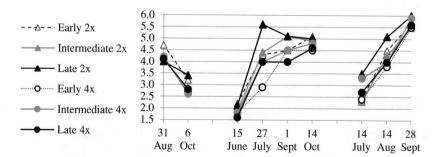

Fig. 2. Brown blight infection of six perennial ryegrass cultivar types in three years. Scale: 1 = no infection, 9 = > 75% covered with symptoms

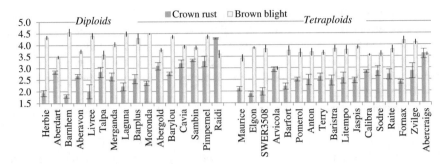

Fig. 3. Perennial ryegrass cultivars ranked in ascending order according to summed infection with two plant pathogens. Means of three years ± standard deviation

taking into account of the entire trial cycle, tetraploids significantly (p < 0.001) approved their superior (by 0.4 pts) resistance. The eight most disease resistant cultivars were tetraploids (Fig. 3). Across the testing period, Barnhem revealed superior resistance against *Pc* (1.8 pts), Arvicola against *Ds* (3.0 pts). Overall tetraploids withstood the two pathogens significantly better than diploids.

4 Conclusion

Varietal effects (V) formed 18–35%, years' effects (Y) 11–26% and V×Y interaction 10–23% of the total variation in disease scores. As a mean of three years, the direct effects of varieties or years on the pathogens' resistance of PRG were highly significant (p < 0.001) in both diploids and tetraploids. The varieties and years exerted significant co-effect onto the occurrence of *Pc* in both ploidy types of PRG and *Ds* in diploid cultivars only. The *Ds* infection in tetraploids did not depend (p > 0.05) on the V×Y interaction. Eight tetraploid cultivars preceded the most resistant diploid on the basis of combined resistance against *Pc* and *Ds*. Most resistant cultivars were LTs. There were 4 LDs and 1 ID among the five highly resistant diploids.

References

Grogan D, Gilliland TJ (2011) A review of perennial ryegrass variety evaluation in Ireland. Ir J Agric Food Res 50:65–81

Hagan A (2005) Leaf spot and rust diseases of turfgrasses. Alabama Cooperative Extension System, ANR-621: 6 p

Kjellström E, Nikulin G, Hansson U, Strandberg G, Ullerstig A (2011) 21st century changes in the European climate: uncertainties derived from an ensemble of regional climate model simulations. Tellus A, Series A 63A(1):24–40

Lancashire JA, Latch GCM (1966) Some effects of crown rust (*Puccinia coronata* Corda) on the growth of two ryegrass varieties in New Zealand. N Z J Agric Res 9:628–640

Pedersen CP (1979) Descriptions of grass diseases no.10. Crown rust. NJF grass disease group: 13–14

Correlation of Seed Yield and Its Components and Chlorophyll Fluorescence Parameters in the Narrow Leafed Lupin (*Lupinus angustifolius* L.)

B. Górynowicz[1(✉)], W. Święcicki[1], W. Pilarczyk[2], and W. Mikulski[1]

[1] Legumes Comparative Genomics Team, Genomics Department,
Institute of Plant Genetics of the Polish Academy of Sciences, Poznań, Poland
bgom@igr.poznan.pl
[2] Department of Mathematical and Statistical Methods,
Poznań University of Life Sciences, Poznań, Poland

Abstract. Narrow leafed lupin (*Lupinus angustifolius* L.) is one of the most important legumes species and a fundamental element for plant and animal production due to high protein content. The aim of this study was to assess the correlation of seed yield and its components on selected chlorophyll fluorescence parameters in narrow leafed lupin cultivars at the flowering and maturation phases. In 2011–2014 field experiments with ten narrow leafed lupin cultivars were conducted at 4 locations and in 2012–2014 chlorophyll fluorescence measurements *in vivo* of plants in field conditions were made. The significant dependence of seed yield on parameters defining the energy flows in terms of cross section and reaction center (ET_o/CS and ET_o/RC) at the end of the flowering phase was found. These parameters were positively correlated with seed yield, number of pods and number of seeds in the pod, and negatively correlated with thousand seed weight. The results of this study suggest the direction of further research connected with the elaboration of breeding method for seed yield using the mobile equipment measuring chlorophyll fluorescence parameters.

Keywords: Chlorophyll fluorescence parameters · *Lupinus angustifolius* L.
Seed yield · Seed yield components

1 Introduction

Narrow leafed lupin (*Lupinus angustifolius* L.) and other lupins are characterized by high potential of seed yield production, especially with reference to protein content. However, the size and quality of seed yield and its components depend on the characteristics of cultivars as well as environmental and agricultural factors.

In breeding programs, the physiological indicators which determine the efficiency of the photosynthetic apparatus and the physiological state are extremely rarely used as selection criteria. However, analysis of chlorophyll fluorescence parameters can be used as a precise tool for assessing the impact of unfavourable environmental factors on

© Springer International Publishing AG, part of Springer Nature 2018
G. Brazauskas et al. (Eds.): *Breeding Grasses and Protein Crops in the Era of Genomics*, pp. 191–195, 2018.
https://doi.org/10.1007/978-3-319-89578-9_35

plants and investigating the response of photosynthesis under stress conditions (Kuckenberg et al. 2009). The aim of this study was to assess the correlation of seed yield and its components on selected chlorophyll fluorescence parameters in narrow leafed lupin cultivars at the flowering and maturation phases.

2 Material and Methods

2.1 Plant Material and Chlorophyll Fluorescence Measurements

Seven traditional and three unbranched cultivars of narrow-leafed lupin (*Lupinus angustifolius* L.) (Górynowicz et al. 2014) from the Polish National List of varieties were used in this work. Field experiments were conducted in randomized complete block design in 2011–2014 at 4 locations: Wiatrowo, Radzików, Karżniczka and Bałcyny. In 2011 the experiments were performed in 4 replicates, while in 2012–2014 6 replicates were applied. From every plot a representative sample of plants from 1 m² was harvested. Seed yield (SY), number of pods (NP), number of seeds per plant (NS), number of seeds per pod (NSPP) and thousand seed weight (TSW) [g] were assessed. In 2012–2014 chlorophyll fluorescence measurements of plants in field conditions were made using the mobile fluorimeter HANDY PEA (*Handy Plant Efficiency Analyser, Hansatech Instruments Ltd., UK*). Measurements were made in each plot on 5 marked plants along each plot, on the most developed leaf from the middle part of the plant, in three periods: at the flowering phase and twice at the green leaf phase after flowering (at the maturation phase). Based on the measured values, the parameters characterizing the energy conversion in photosystem II (PSII) (Kalaji and Łoboda 2010) were calculated using the fluorescent JIP test (Lazár 1999; Lazár and Pospišil 1999; Strasser et al. 2000).

2.2 Statistical Analysis

The chlorophyll fluorescence parameters (chflp) results were analysed statistically independently for each year using the analysis of variance (ANOVA) under a linear model for randomized complete block design (Elandt 1964) and for a series of experiments under a mixed model usually used in such analysis (Cochran and Cox 1950). The analysis of correlations, the analysis of multiple regression (Draper and Smith 1973) and the accumulated analysis of variance was applied in order to identify relationships between SY and its components and chflp, using the MS EXCEL 2016 program and the GENSTAT package (Payne et al. 1987). The coefficients of determination R^2 were calculated and the regression coefficients a_1 with their empirical significance levels were computed.

3 Results and Discussion

The analysis of variance within each measurement year showed, that narrow leafed lupin cultivars were best differentiated at the green leaf phase after flowering (second and third periods) – significant differences at the level $\alpha = 0.05$. At the flowering phase, the significant differences at the $\alpha = 0.01$ level were observed. The analysis of series of experiments confirmed the differences among cultivars, especially at the green leaf phase. Based on the level of variation of cultivars in terms of chflp (the analysis of variance in first measurement year for three periods) and the correlation between these parameters in the first year for three periods, 5 chflp (Fv/Fm, psi0, P.I.csm, ET_o/CS, ET_o/RC) (Kalaji and Łoboda 2010) were selected for further research. Based on the analysis of correlation between SY and its components and selected chflp, an average from measurements and research years (Table 1), selected chflp were divided into two groups. Parameters defining the energy flows in terms of cross section and reaction center (ET_o/CS and ET_o/RC), which positively correlated with SY, NP and NSPP, and negatively correlated with TSW were assigned to the first group. The second group formed other parameters (Fv/Fm, psi0, P.I.csm), which negatively correlated with SY and TSW and generally positively correlated with NP and NSPP (Table 1).

The analysis of multiple regression for relationship of SY and selected chflp average from measurements and research years, showed statistically highly significant ($\alpha < 0.01$) dependence of SY on ET_o/CS and ET_o/RC. The results of the analysis of multiple regression for relationship between SY and ET_o/CS and ET_o/RC (on average from each period and research years), showed statistically highly significant dependence of SY on these parameters in the second period (at the beginning of the green leaf phase after flowering). The coefficient of determination R^2 for this relationship was very high and reached 60,5%. In the accumulated analysis of variance for relationship between SY and these parameters in the second period, average from research years and cultivars, empirical level of significance was $\alpha < 0.001$ (Table 2). For this relationship considering all research years, the coefficient R^2 accounted for 51.6% and the regression coefficients a_1 were statistically highly significant. However, for relationship considering each year, the coefficient R^2 accounted for 60.4% but the regression coefficients were statistically insignificant ($\alpha = 0.05$). SY depend on chflp, especially on ET_o/CS and ET_o/RC in the second period. Measurements of PSII photochemical activity showed a general decrease of ET_o/CS and ET_o/RC, indicating the worse condition of the narrow leafed lupin leaf at the end of the vegetation season (Kalaji and Łoboda 2010; Kalaji et al. 2016).

Table 1. Coefficients of correlation between seed yield and its components and selected chlorophyll fluorescence parameters averaged over experiment years.

Trait	chflp				
	Fv/Fm	psi0	P.I.csm	ET_o/CS	ET_o/RC
SY	−0,17	−0,13	−0,13	0,10	0,30
NP	−0,14	0,30	0,25	0,26	0,08
NSPP	0,50	0,16	0,31	0,60	0,45
TSW	−0,59	−0,31	−0,50	−0,64	−0,28

chflp – chlorophyll fluorescence parameters, SY – seed yield, NP – number of pods, NSPP – number of seeds per pod, TSW – thousand seed weight

Table 2. Accumulated analysis of variance for relationship between seed yield and ET_o/CS and ET_o/RC averaged over years and cultivars (excluding cvs. Boruta and Graf).

Source of variation	Degrees of freedom	Sum of squares	Average square	Empirical F	Empirical level of significance
ET_o/CS	1	67.2357	67.2357	173.22	<0.001
ET_o/RC	1	5.9150	5.9150	15.24	<0.001
Year	2	12.9355	6.4678	16.66	<0.001
Error	139	53.9530	0.3882		
Total	143	140.0392	0.9793		

4 Conclusion

The significant correlation between seed yield and parameters defining the energy flows in terms of cross section and reaction center at the beginning of the green leaf phase after flowering suggests the direction of further research connected with the elaboration of breeding method for seed yield using the mobile equipment measuring chlorophyll fluorescence parameters. Estimating the chlorophyll fluorescence parameters may be useful for finding and sorting cultivars tolerant against environmental conditions.

Acknowledgments. This study was funded by the National Multi-Year Project implemented by the Ministry of Agriculture and Rural Development in Poland.

References

Cochran WG, Cox GM (1950) Analysis of the results of a series experiments. In: Experimental designs. Wiley, New York, pp 545–568

Draper NR, Smith H (1973) Analiza regresji stosowana. Państwowe Wydawnictwo Naukowe, Warszawa

Elandt R (1964) Statystyka matematyczna w zastosowaniu do doświadczalnictwa rolniczego. Państwowe Wydawnictwo Naukowe, Warszawa, pp 360–361

Górynowicz B, Święcicki W, Osiecka A, Kaczmarek Z (2014) Terminal inflorescence and restricted branching genes in lupin (L. *albus* L., *L. angustofolius* L., L. *luteus* L.) and field bean (*Vicia faba* L.) breeding in Poland. J Agric Sci Technol B 4:702–711

Kalaji MH, Łoboda T (2010) Fluorescencja chlorofilu w badaniach stanu fizjologicznego roślin. Wydawnictwo SGGW, Warszawa

Kalaji MH, Center MD, Dąbrowski P, Samborska IA, Łukasik I, Swoczyna T, Pietkiewicz S, Bąba W (2016) Zastosowanie pomiarów fluorescencji chlorofilu w badaniach środowiskowych. Kosmos 65(2):197–205

Kuckenberg J, Tartachnyk I, Noga G (2009) Temporal and spatial changes of chlorophyll fluorescence as a basis for early and practise detection of leaf rust and powdery mildew infections in wheat leaves. Precis Agric 10:34–44

Lazár D (1999) Chlorophyll a fluorescence induction. Biochemica et Biophysica Acta 1412:1–28

Lazár D, Pospišil P (1999) Mathematical simulation of chlorophyll a fluorescence rise measured with 3-(3′,4′-dichlorophenyl)-1,1 dimetylurea-treated barley leaves at room and high temperatures. Eur Biophys J 28:468–477

Payne RW, Lane PW, Ainsley AE, Bricknell KE, Digby PGN, Harding SA, Leech PK, Sampson HR, Todd AD, Verrier PJ, White RB (1987) GENSTAT 5 Reference Manual. Clarendon Press, Oxford

Strasser RJ, Srivastava A, Tsimilli-Michael M (2000) The fluorescent transient as a tool to characterize and screen photosynthetic samples. In: Yunus M, Pathre U, Mohanty P (eds) Probing photosynthesis: mechanisms, regulation and adaptation. Taylor and Francis, London, pp 445–483

Seventh Workshop of the EUCARPIA Festulolium Working Group

M. Ghesquière[✉]

INRA/URP3F, Le Chêne, RD 150, CS 80006, 86600 Lusignan, France
Marc.Ghesquiere@inra.fr

Abstract. Following the last meeting in Olomouc in 2016, the 7th workshop of the FWG was held within the timeframe of the joint meeting of the Fodder Crops and Amenity Grasses Section and the Oil and Protein Crops Section of EUCARPIA. About 30 people attended the workshop which started by three presentations: (i) A history of Festulolium in Poland, by Prof Z. Zwierzykowski; (ii) A sure-root to climate-smart grassland, by Prof M.W. Humphreys; (iii) Current Festulolium breeding at ILVO, by Dr J. Baert. The FWG warmly thanked their colleagues for comprehensive presentations, especially M.W. Humphreys and Z. Zwierzykowski in the view of their forthcoming retirement. As such, the scope of their presentations surely exceeds that of the specialist workshop audience and the FWG encouraged them to communicate and disseminate their presentation by any means they consider appropriate. Then, two topics were further debated on the conservation and the dissemination of Festulolium plant material and varieties as genetic resources. The multisite field trial undertaken in 2012 is nearly finished, practical issues about it were considered as well as the development of new common experimental work.

Keywords: Festulolium · Genetic resources · Variety testing

1 Multilocation Festulolium Field Trial

The trial launched in 2012 is now complete at all locations including the latest trials begun in or after 2012 (Teagasc and ILVO). Overall, it includes comparisons of yield from 21 varieties comprising 15 Festulolium cvs and their parental controls, over three years at all locations. Various other traits such as disease susceptibility were scored additionally as well as chemical compositions at 4 locations. It is proposed that Vladimir Černoch who collated the data from the beginning and circulated them to participants on an annual basis, finalizes the data base and compiles all within one Excel spreadsheet and forwards a copy by the end of the year to all those who participated to the trial within the FWG. The use of the database and further communication of the results will be left to the freedom of each partner. A good way to indicate that all was achieved through activities undertaken by the FWG would be to quote one of the collective presentations at Eucarpia (Ghent, 2015 or Vilnius, 2017). The collective scientific paper including GISH and DArT markers from David Kopecký is well advanced and should be submitted by the end of the year.

© Springer International Publishing AG, part of Springer Nature 2018
G. Brazauskas et al. (Eds.): *Breeding Grasses and Protein Crops in the Era of Genomics*, pp. 196–200, 2018.
https://doi.org/10.1007/978-3-319-89578-9_36

2 New Common Work

No further common experiment was considered yet at this FWG although ideas and opportunities for future consideration are welcome from all within the FWG. It is obvious that new genotyping facilities could include individual plant phenotype assessment with prospects for GWAS and GS. Evaluation of Festulolium cvs in a mixture with legumes as it was evoked in Olomouc last year would be certainly a useful field of common work. Mike Humphreys also underlined in his presentation the potential of Festulolium to include ecosystem services in addition to usual forage targets. M. Humphreys described two such projects in his presentation (SureRoot and Climate Smart Grasslands). He also described opportunities for EU participants in the FWG to join IBERS in seeking funding for projects with Chinese institutes (Shanxi Agricultural University (SAU) and Beijing Academy of Agriculture and Forestry Sciences (BAAFS)). BAAFS have acquired significant funding for grassland projects on drought resistance and for land restoration and welcome European partners. M. Humphreys invited any member of the FWG to contact him if they are interested to develop ideas for these or other possible areas of joint research. As he will soon retire he suggests further contacts should include his IBERS colleague Dr John Scullion (jos@aber.ac.uk) who will take over from him the BBSRC-funded UK-China institute partnership.

3 Genetic Resources

This is an important issue to enable more research in the future on Festulolium as an object for both science and breeding. Liv Østrem and Marc Ghesquière will update the list of Festulolium varieties collected from various sources in 2009, and check for content by circulating within the FWG (Table 1). M. Ghesquière will contact Evelin Willner at IPK in Malchow to update the current status of Festulolium, either as spontaneous accession if any, or as old/removed of EU lists bred varieties. E. Willner agreed to carry out this undertaking by requesting through EURISCO. If at all possible, the situation of potential parent species of Festulolium such as *F. pratensis* var *apennina*, *F. arundinacea* var *glaucescens* and *F. mairei* will be documented. A further objective for the FWG towards the gene banks is to finalize some written documents/protocols in respect of specific requirements for developing Festulolium: passport/genealogy data, size and germination rate of seed sample, seed regeneration rate and procedures undertaken. The FWG expressed thanks to E. Willner for her help. It was suggested the process could be further supported through future support and agreement with other Gene Banks such as that in Aberystwyth or the Nordic Gene Bank.

198   M. Ghesquière

Table 1. List of the Festulolium cvs officially registered in one national list, at least, of the EU countries in 2017 and according maintainer.

Cv denomination	National list[a]	Maintainer		
		N°	Company	Address
Aberniche	UK	-	Germinal Holdings Ltd	2 Commercial Rd, Banbridge BT32 3ET · UK
Achilles	CZ DE FR SK	CZ 201	DLF Seeds	Fulnecká 95 Hladké Životice 742 47 CZ · CZ
Agula	PL	PL 153	Danko Hodowla Roślin sp. z o.o.	Choryń 27 Kościan 64-000 PL · PL
Bečva	IT	IT 918	DLF Seeds	Fulnecká 95 Hladké Životice 742 47 CZ · CZ
Fabel	NO	NO 499	Graminor AS	Hommelstadvegen 60 Ridabu 2322 NO · NO
Fedoro	DE	DE 39	Deutsche Saatveredelung AG	Postfach 1407 Lippstadt 59557 DE · DE
Felina	CZ EE FI LT LV NO SK	CZ 201	DLF Seeds	Fulnecká 95 Hladké Životice 742 47 CZ · CZ
Felopa	PL DE DK	PL 153	Danko Hodowla Roślin sp. z o.o.	Choryń 27 Kościan PL · PL
Felovia	CH	CH 162	Agroscope ART	Reckenholzstrasse 191 Zürich 8046 CH · CH
Festilo	BE	BE 218	ILVO Plant Toegepaste Genetica en Veredeling	Caritasstraat, 21 Melle 9090 BE · BE
Festum	IT	IT 1324	Natura s.l.r.	Via d'Annunzio, 207, Pineto Teramo 64025 IT · IT
Fojtan	CZ	CZ 201	DLF Seeds	Fulnecká 95 Hladké Životice 742 47 CZ · CZ
Frosta	NO	NO 499	Graminor AS	Hommelstadvegen 60 Ridabu 2322 NO · NO
Hathor	CZ	CZ 201	DLF Seeds	Fulnecká 95 Hladké Životice 742 47 CZ · CZ
Helus	CZ	CZ 201	DLF Seeds	Fulnecká 95 Hladké Životice 742 47 CZ · CZ
Hemsut	CZ	CZ 201	DLF Seeds	Fulnecká 95 Hladké Životice 742 47 CZ · CZ
Hermes	CZ	CZ 201	DLF Seeds	Fulnecká 95 Hladké Životice 742 47 CZ · CZ
Hipast	CZ DK FR	CZ 201	DLF Seeds	Fulnecká 95 Hladké Životice 742 47 CZ · CZ
Honak	CZ	CZ 201	DLF Seeds	Fulnecká 95 Hladké Životice 742 47 CZ · CZ

(*continued*)

Table 1. (*continued*)

Cv denomination	National list[a]	Maintainer			
		N°	Company	Address	
Honor	CZ	CZ 201	DLF Seeds	Fulnecká 95 Hladké Životice 742 47 CZ	CZ
Horimir	CZ	CZ 201	DLF Seeds	Fulnecká 95 Hladké Životice 742 47 CZ	CZ
Hostyn	CZ PL SK	CZ 201	DLF Seeds	Fulnecká 95 Hladké Životice 742 47 CZ	CZ
Hyperon	CZ	CZ 201	DLF Seeds	Fulnecká 95 Hladké Životice 742 47 CZ	CZ
Kebo	CZ	CZ 201	DLF Seeds	Fulnecká 95 Hladké Životice 742 47 CZ	CZ
Lenor	SK	SK 629	Graminex, s.r.o.	Okruzna 1555/13 Levoca 054 01 SK	SK
Lifema	FR NO	FR S8233	Deutsche Saatveredelung AG	Postfach 1407 Lippstadt 59557 DE	DE
Lina DS	LT	LT 32	LAMMC	Instituto a. 1, Akademija, Kedainiai distr. 58344 LT	LT
Lofa	CZ EE HR SK RO	CZ 201	DLF Seeds	Fulnecká 95 Hladké Životice 742 47 CZ	CZ
Lueur	FR ·	FR 9504	Agri-Obtentions	Chemin de la Petite Minière BP 36 Guyancourt Cédex 78041 FR	FR
Mahulena	PL DE	PL 1035	DLF Seeds	Fulnecká 95 Hladké Životice 742 47 CZ	CZ
Matrix	IT	IT 282	Cropmark Seeds Ltd.	PO Box 16 574 Christchurch NZ	NZ
Merlin	SI	SI 502	INIA	La Estanzuela, C. Correo 39173 Colonia 70000 UY	UY
Paulita	NO	NO 29	DLF Trifolium	Postboks 19 Store Heddinge 4660 DK	DK
Perseus	CZ DE SK UK	CZ 201	DLF Seeds	Fulnecká 95 Hladké Životice 742 47 CZ	CZ
Perun	CZ EE HR SK	CZ 201	DLF Seeds	Fulnecká 95 Hladké Životice 742 47 CZ	CZ
Puga	LT	LT 32	LAMMC	Instituto a. 1, Akademija, Kedainiai distr. 58344 LT	LT
Punia DS	LT	LT 32	LAMMC	Instituto a. 1, Akademija, Kedainiai distr. 58344 LT	LT
Rebab	CZ	CZ 201	DLF Seeds	Fulnecká 95 Hladké Životice 742 47 CZ	CZ
Revolution	IT	IT 282	Cropmark Seeds Ltd.	PO Box 16 574 Christchurch NZ	NZ

(*continued*)

Table 1. (*continued*)

Cv denomination	National list[a]	Maintainer		
		N°	Company	Address
Spring Green	IT	IT 777	Pure Seed Testing Inc.	West Street 73G Hubbard US	US
Sulino	PL	PL 153	Danko Hodowla Roślin sp. z o.o.	Choryń 27 Kościan 64-000 PL	PL
Vizule	LV	LV 189	Latvia University of Agriculture	Liela Street 2 Jelgava 3001 LV	LV
Vétra	LT	LT 32	LAMMC	Instituto a. 1, Akademija, Kedainiai distr. 58344 LT	LT

[a]International country 2-letters code

4   Evaluation of the Varieties and the Taxonomy of the Hybrids

Official testing is also an important issue for fair assessment, registration and recommendation in the EU lists. There is a window to improve this thanks to the contact of DLF (V. Černoch) by the officials of The NL. V. Černoch will circulate their proposal within the FWG so that everyone can add comments and further suggestions. One objective would be to agree with a common protocol which could be proposed on behalf of the FWG to any official testing in EU having to assess Festulolium candidates in the next years. On a taxonomy level, D. Kopecký has contacts in Italy in order to make more rational the botanical names of the hybrid combinations behind the Festulolium cvs as well as to propose varieties which could be hold as true-to-type. D. Kopecký will circulate his proposition to the FWG in this way to be transmitted then to the group of taxonomists.

5   The Festulolium Working Group Core Group

The core group is presently composed of: L. Østrem, V. Černoch, M. Ghesquière, M. Humphreys and Z. Zwierzykowski. M. Ghesquière replaced M. Humphreys as the head of the core group in 2016; Furthermore, M. Humphreys proposed that Danny Thorogood replaces him in the core group (for UK) in the next year. In the same way, Z. Zwierzykowski will propose a new person to replace him. Of course, both M. Humphreys and Z. Zwierzykowski remain welcome to attend the next meeting of the FWG. D. Kopecký suggested that Vilma Kemešytė joins the core group due to her active work in Festulolium breeding at the Lithuanian Research Centre for Agriculture and Forestry. As the Section of EUCARPIA will skip the meeting in 2018, the question of meeting the FWG next year remains open; any suggestion in this way is welcome. In any case, the FWG will meet at the EUCARPIA meeting in Zurich in 2019.

Implementation of Phenomic and Genomic Tools in Breeding

Using LIDAR for Forage Yield Measurement of Perennial Ryegrass (*Lolium perenne* L.) Field Plots

K. Ghamkhar[1]([✉]), K. Irie[2], M. Hagedorn[2], J. Hsiao[2], J. Fourie[1],
S. Gebbie[3], C. Flay[1], B. Barrett[1], A. Stewart[4], and A. Werner[2]

[1] Forage Science, Grasslands Research Centre, AgResearch,
Palmerston North, New Zealand
kioumars.ghamkhar@agresearch.co.nz
[2] Lincoln Agritech Ltd, Lincoln, New Zealand
[3] Development Engineering, Lincoln Research Centre, AgResearch,
Lincoln, New Zealand
[4] PGG Wrightson Seeds, Christchurch, New Zealand

Abstract. Perennial ryegrass (*Lolium perenne* L.) is the most important forage species in New Zealand and one of the most important species in Europe and Australia. A LIDAR-based tool for non-invasive estimation of plant biomass in perennial ryegrass field plots was developed with the aim to measure the variation for plant foliage yield among 12 cultivars. Phenotypic data were collected both with a mechanical harvester and using the LIDAR system over the growing season. Significant ($P < 0.05$) variation was observed among cultivars. Significant ($P < 0.0005$) correlations were observed between processed LIDAR data and fresh and dry weights of plant foliage biomass with R^2 values of 0.78 and 0.76, respectively.

Keywords: *Lolium perenne* · Biomass · Foliage yield · Growth rate
LIDAR

1 Introduction

Perennial ryegrass (*Lolium perenne* L.) is the most widely used temperate grass (Charlton and Stewart 1999) and generates $14 billion of value for New Zealand's economy each year (Nixon 2015). However, its foliage yield can vary widely depending on factors such as cultivar, soil, environment, fertilisation, cultivation time, pest management and, indirectly, grazing time and intensity and also grower's knowledge of all aforementioned factors.

To accelerate the rate of yield gain, and increase the efficiency of methods such as genomic selection, the focus is to divert some of the time and cost lost for yield measurement into high throughput and repeatable yield measurement. Improving the quality of foliage yield measurement is viewed as a means of increasing the effect of genomics-assisted selection because forage productivity may be enhanced with less information lost due to increased accuracy and precision in yield measurement (Pollock et al. 2005). The aim of this study was to measure the variation for plant foliage yield

© Springer International Publishing AG, part of Springer Nature 2018
G. Brazauskas et al. (Eds.): *Breeding Grasses and Protein Crops in the Era of Genomics*, pp. 203–208, 2018.
https://doi.org/10.1007/978-3-319-89578-9_37

among 12 cultivars of perennial ryegrass using a LIDAR-based system and to evaluate growth rate in these cultivars in comparison with manual biomass measurement.

2 Materials and Methods

A prototype machine was designed to enable LIDAR data collection, and algorithms were developed for data processing. An algorithm was developed for grass plot segmentation, ground surface detection, and estimation of plant biomass.

A total of 36 rows of perennial ryegrass, each 2 m long, comprising three replicates of 12 cultivars (Table 1) were grown in a random order and distribution among 108 rows of perennial ryegrass populations with 50 cm side space between rows at the PGG Wrightson's Kimihia site (−43.62, 172.46) near Lincoln, New Zealand in March 2015. These rows were clearly marked. After 41 days, these rows were scanned using the prototype machine at three time points at early, mid and late time points during May to June 2015; 03/05, 17/05 and 14/06, with a mechanical harvest and yield data collection at the final scanning date. This was done to measure dry matter yield and also fresh weight at these time points. The vegetative parts including leaf, and tillers were cut immediately after the last screening and dried for dry matter determination. Coefficient of variation ($CV = \sigma/\mu$) of dry matter yield, fresh weight and scanning measurements among cultivars where σ is standard deviation and μ is sample mean was estimated in Excel. The calculated biomass estimates were validated with regression analysis against harvest data. Correlation between harvest data and LIDAR reads at the last measurement point were calculated.

Table 1. Cultivars of perennial ryegrass examined in this study, their symbiont endophytes, their ploidy level and species categorisation.

Cultivar	Endophyte	Ploidy level	Species category
One50	AR37	2x	Perennial ryegrass
BanquetII	Endo5	4x	Hybrid ryegrass
Base	AR37	4x	Perennial ryegrass
Excess	AR37	2x	Perennial ryegrass
Expo	AR37	2x	Perennial ryegrass
Ohau	AR37	4x	Hybrid ryegrass
Prospect	AR37	2x	Hybrid ryegrass
Rely	AR37	2x	Perennial ryegrass
Request	AR37	2x	Perennial ryegrass
Reward	Endo5	4x	Perennial ryegrass
Samson	AR37	2x	Perennial ryegrass
Victorian	Standard	2x	Perennial ryegrass

3 Results and Discussion

Observed variation among cultivars for all yield traits were low to moderate (Table 2). While the lowest variation was observed in % dry matter, the highest variation among cultivars was in fresh weight. Variation for the LIDAR reads was very close to variation for fresh and dry weighs, and specifically dry weight, suggesting relative precision of these reads for yield measurement.

Table 2. Variation for foliage yield data among 12 cultivars of perennial ryegrass obtained at the last measurement date.

Trait	Mean	Standard deviation	CV%
Fresh Weight (FW), g	479.47	94.57	19.72
Dry Weight (DW), g	150.88	25.04	16.59
Dry Matter (DM), %	31	2.26	7.13
LIDAR volumetric read	18935774.83	2923314.65	15.43

Point to point correlation between LIDAR data and fresh weight (Fig. 1a) and dry weight (Fig. 1b) showed strong linear association between LIDAR data and both these measures of forage yield; $R^2 = 0.79$ and 0.76, respectively (with p values of 4.7×10^{-30} and 1.3×10^{-27}). Estimation of %DM, however, was less than convincing with no linear correlation and a feeble polynomial correlation at $R^2 = 0.35$ (Fig. 2), $y = 61.344x^4 - 5441x^3 + 119288x^2 - 44583x + 1E + 07$ where y is LIDAR estimate and x is %DM.

Significant variance in forage yield was observed among the 12 cultivars ($P = 0.05$) (Fig. 3). Cultivar Base (4x and inoculated with endophyte AR37) was the highest yielding cultivar whereas cultivars Victorian (2x and inoculated with standard endophyte), Prospect (hybrid), Excess and One 50 (all 2x and inoculated with endophyte AR37) were the lowest yielding cultivars. It was noted that the forage yield or biomass is not only dependent on the cultivar, but also highly depended on the location of the segment. Also, while tetraploid cultivars yield more biomass, a good combination of diploid grass and endophyte can equally result in higher yields. The results of our study indicate that volumetric LIDAR data is a good indicator of fresh and dry weights in perennial ryegrass.

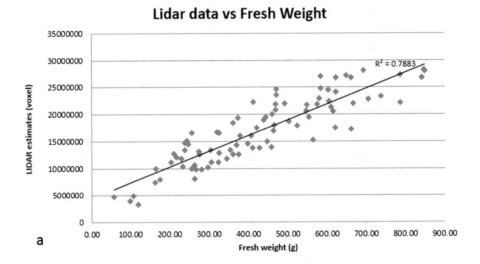

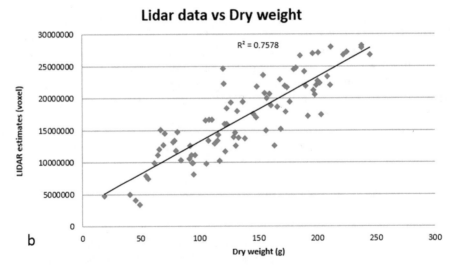

Fig. 1. Correlation between LIDAR volumetric estimates (LIDAR reads) and fresh (a) and dry (b) weights in 12 cultivars of perennial ryegrass.

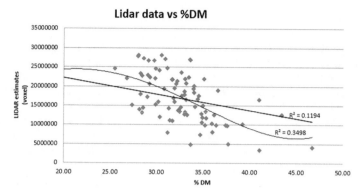

Fig. 2. Correlation between LIDAR volumetric estimates (LIDAR reads) and % dry matter (%DM) in 12 cultivars of perennial ryegrass.

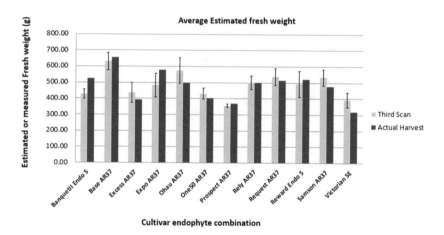

Fig. 3. Comparison of yield estimation from LIDAR volumetric data (light colored) and measured fresh weight obtained from actual harvest (dark colored) in 12 cultivars of perennial ryegrass. Error bars represent the standard deviation of the absolute difference between the predicted and actual fresh weight in each cultivar.

4   Conclusion

LIDAR volumetric data is a good indicator of fresh and dry weight in perennial ryegrass regardless of the stage of growth and time of measurement. While ploidy level directly correlates with forage yield, a combination of diploid cultivars and right endophyte can equally result in high forage yield.

Acknowledgments. This study was funded by "Pastoral Genomics Plus" implemented by AgResearch, PGG Wrightson Seeds and Lincoln Agritech Ltd.

References

Charlton JF, Stewart AV (1999) Pasture species and cultivars used in New Zealand-a list. In: Proceedings of the conference-New Zealand Grassland Association, pp 147–166

Nixon C (2015) How valuable is that plant species? Application of a method for enumerating the contribution of selected plant species to New Zealand's GDP. Report New Zealand Institute for Economic Research

Pollock C, Abberton M, Humphreys M, McGilloway D (2005) Grass and forage improvement: temperate forages. Grassland: a global resource. Plenary and invited papers from the XX International Grassland Congress, Dublin, Ireland, Wageningen Academic Publishers, pp 57–68

An Indoor Screening Method for Reduced Fall Dormancy in Alfalfa

A. Bertrand(✉), A. Claessens, and S. Rocher

Quebec Research and Development Centre, Agriculture and Agri-Food Canada,
Québec, QC G1V 2J3, Canada
Annick.bertrand@agr.gc.ca

Abstract. Fall dormancy (FD) is a determinant component of alfalfa yield in northern climate, but selection for that trait is often done at the expense of freezing tolerance. Field selection for low dormancy is a tedious process that depends on unpredictable fall conditions to identify less dormant genotypes. Here we describe two screening methods for low dormancy in alfalfa which can be entirely performed indoor. Both methods rely on plant exposure to a critical photoperiod during four weeks to discriminate plants according to their FD level. One method also includes successive freezing stress to simultaneously improve freezing tolerance (TF) and FD. These approaches can be integrated in a recurrent selection program to produce populations combining reduced FD and high TF.

Keywords: *Medicago saliva* L. · Dormancy · Freezing tolerance
Recurrent selection

1 Introduction

Breeding for low FD is an efficient approach to increase the annual yield of alfalfa by extending its growing seasons from late summer through early winter. Less dormant cultivars, however, have lower winter survival than more dormant cultivars. While the phenotypic correlation between these two traits is high (Stout and Hall 1989; Schwab et al. 1996), the genotypic correlation is low, suggesting that FD and winter survival can be improved simultaneously (Brummer and Casler 2014).

FD rating of alfalfa cultivars includes eleven classes broadly distributed into three subgroups: fall dormant (FD 1–4), semi-dormant (FD 5–7) and non-dormant (FD 8–11) (Ariss and Vandemark 2007). There is a large genetic diversity for that trait that translated into successful selection for reduced dormancy within cultivars with various initial dormancy levels (Cunningham et al. 2001). The level of FD in alfalfa is usually determined using a standard FD test based on shoot height, measured approximately 25 days following the last cut in the fall, and compared to the height of cultivars of known FD classes (Teuber et al. 1998). The validity of this test, that could only be performed once a year in the field, depends largely on unpredictable fall conditions. For instance, regrowth has to occur during late fall but before a severe frost as low temperatures after the last cut in the fall could inhibit plant fall regrowth and invalidate the test. In Nordic regions experiencing harsh winter climate, two years are often necessary to establish

© Springer International Publishing AG, part of Springer Nature 2018
G. Brazauskas et al. (Eds.): *Breeding Grasses and Protein Crops in the Era of Genomics*, pp. 209–214, 2018.
https://doi.org/10.1007/978-3-319-89578-9_38

reliable FD rating. As a result, costly assessment of the genetic material at multiple locations over many years is often needed to accurately discriminate plants regarding to their FD level. The unpredictability of field tests is expected to be exacerbated by climate change, as warmer temperatures in the fall will be concomitant to short photoperiods (Dalmannsdottir et al. 2016). New approaches are needed by breeding programs to accelerate and reduce the cost of FD assessment. For instance, the development of an indoor test would give the possibility to combine freezing test and FD test and to improve both traits simultaneously. Our objectives were to (1) develop an indoor test for FD assessment; (2) develop an indoor selection method for reduced FD; and (3) develop an indoor selection method for simultaneous improvement of FD and freezing tolerance.

2   Materials and Methods

1. Dormancy test performed indoor under controlled conditions

The FD test was developed using plants of semi-dormant cultivar 6010 (FD6) exposed to four photoperiods. To determine the optimal photoperiod to discriminate between FD levels, 1200 plants of alfalfa cultivar 6010 were grown for four weeks in a growth chamber under 16 h photoperiod and 22/17 °C day/night (d/n) temperature. Two successive cuts separated by four weeks of regrowth were performed to homogenize plants development stage. Plants were then transferred into four growth chambers (300 plants per growth chamber) and exposed to different photoperiods (8, 10, 12 or 14 h photoperiod) with 18/15 °C d/n temperature regime. After three weeks, the critical photoperiod accurately discriminating between dormant and less dormant genotypes was determined when 50% of the plants were remaining dormant. To validate the critical photoperiod determined experimentally, three cultivars with known FD level were included in the test for validation of FD rating: 6010 (FD 6), a cultivar that we are currently selecting for reduced dormancy and increased freezing tolerance, CUF 101(FD 9), a check cultivar less dormant than 6010 identified in Teuber et al. 1998, and Apica (FD 4), a cultivar more dormant than 6010. One year-old field-grown plants of these cultivars were transplanted in pots (40 pots of 10 plants x three cultivars). Plants were cut and pots were transferred into a large growth chamber and exposed to the critical photoperiod of 12 h previously determined and 18/15 °C d/n temperature. Plant height was measured after 15 and 25 days of regrowth under these conditions.

2. Indoor selection method for reduced FD

We developed a selection protocol to improve FD in freezing tolerant material (Fig. 1). Briefly, 2000 plants from an alfalfa cv. Yellowhead (FD2) were grown in individual tubes for four weeks. Plants were then cut and exposed to a 14-h photoperiod which we previously determined to discriminate between dormant and non-dormant genotypes in alfalfa cultivars originating from higher latitudes such as Yellowhead. Plants were exposed to three cycles of growth/cut/regrowth. After each cycle, the more dormant genotypes were eliminated (around 70% at each cycle), to

SELECTION FOR REDUCED DORMANCY (RD)

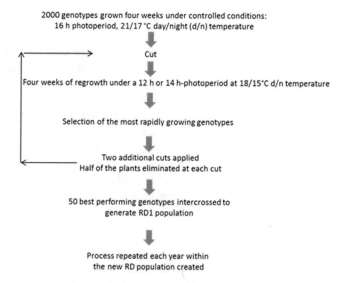

Fig. 1. Schematic illustration of the procedure used for selection of populations with reduced dormancy.

finally end-up with the 50 tallest genotypes that were selected and intercrossed to generate a new population with reduced dormancy (RD+1).

3. Indoor selection method for reduced FD and increased freezing tolerance

We previously developed an indoor recurrent phenotypic selection test to improve freezing tolerance in alfalfa populations (Bertrand et al. 2014). Using this approach, several cycles of recurrent selection were performed in six alfalfa cultivars, allowing a progressive increase of adaptive alleles frequency and promoting epistatic interactions in the recurrently selected populations. Briefly, three successive freezing stresses were applied on a large number of genotypes from one initial alfalfa genetic background (cv. 6010) to progressively select the hardiest plants. At the completion of the selection process, the 10% superior genotypes were intercrossed to generate a new population with improved tolerance to freezing (TF populations). To simultaneously reduce FD and increase freezing tolerance, we added a step of selection for reduced FD to this indoor selection method for freezing tolerance. Therefore, after each freezing stress, plants were exposed to a 12 h-critical photoperiod under 18/15 °C (d/n) for the 4-weeks regrowth period, resulting in three cycles of joint selection for freezing tolerance and FD. The most rapidly growing genotypes that survived each freezing stress were selected and, after three cycles, the 50 best performing genotypes were intercrossed to generate a new population with both reduced FD and improved freezing tolerance.

3 Results and Discussion

3.1 FD Test Performed Indoor

The proportion of genotypes of semi-dormant cultivar 6010 remaining dormant after an exposure to 8, 10, 12 and 14 h-photoperiod was respectively of 82%, 61% 50% and 8% (Table 1). This test revealed 12-h as a critical photoperiod at which half of the plants start to regrow while half of the plants remain dormant. Therefore, this photoperiod was selected to perform our indoor FD test.

We applied this test to alfalfa cultivars of contrasting FD classes (CUF 101: FD 9, 6010: FD 6, and Apica: FD 4). Significant regrowth differences were observed between cultivars after exposure to a 12-h photoperiod for 15 or 25 days (Fig. 2). However, a better discrimination between cultivars was obtained after 25 days of regrowth, which was chosen as the standard regrowth period in our indoor test. A 25-days regrowth was also recommended in standard FD test in the field (Teuber et al. 1998).

Table 1. Effect of photoperiod on dormancy levels of alfalfa cv. 6010. The total number of plants, the number of dormant and non-dormant plants, and the proportion of plant remaining dormant after 25 days of exposure to four different photoperiods (8, 10, 12 or 14 h) are indicated.

Photoperiod tested	8 h	10 h	12 h	14 h
Total number of plants	293	299	299	299
Number of non-dormant plants	52	117	148	274
Number of dormant plants	241	182	151	25
% dormant plants	82.3	60.9	50.5	8.4

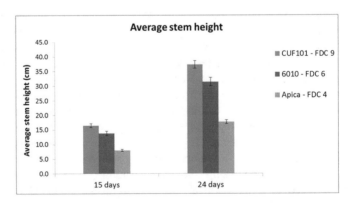

Fig. 2. Average stem height of alfalfa cultivar CUF 101 (FD 9), 6010 (FD 6), and Apica (FD 4), measured after 15 and 25 days of regrowth under a 12 h photoperiod and a 18/15 °C d/n temperature regime.

3.2   Indoor Selection Method for Reduced FD

We identified less dormant alfalfa genotypes within a cultivar of FD class 2 and developed a population with reduced FD by applying one cycle of selection (Fig. 3). The selection method could be optimized to select for reduced FD within cultivars from various origins by evaluating the cultivar-specific critical photoperiod.

Fig. 3. Visual assessment of regrowth of alfalfa cv. Yellowhead (FD2) after one cycle of indoor selection for reduced dormancy. Less dormant genotypes are in tray A while more dormant genotypes are in tray B. The genotypes in tray A will be intercrossed to generate RD+1 population.

3.3   Indoor Selection Method for Reduced Dormancy and Increased Freezing Tolerance – A Perspective

Plants from two semi-dormant (FD5 and FD6) alfalfa cultivars were grown in individual tubes for four weeks and we applied our 'Indoor selection method for reduced FD and increased freezing tolerance'. After three cycles of growth/freezing stress/regrowth under a discriminant photoperiod of 12 h, the 50 tallest genotypes were selected and intercrossed to generate a new population with both improved freezing tolerance (TF+1) and reduced dormancy (RD+1). Seeds were collected and a second cycle of selection was performed within both backgrounds. The improved freezing-tolerant-reduced dormancy (TF-RD+2) populations will soon be evaluated for their winter survival and fall dormancy level in the field.

4   Conclusion

To address issues associated with the unpredictability and resource intensive assessment of FD in the field, we developed indoor selection protocols that were used to select within alfalfa cultivars of different FD classes. Our test confirmed previous reports showing that fall growth and winter hardiness can be improved simultaneously (Li et al. 2015). These indoor tests will be particularly useful to identify genotypes that combine superior winter hardiness and low FD within alfalfa breeding material. The identification of these genotypes is essential to develop breeding populations and accelerate the development of cultivar adapted to northern areas.

References

Ariss JJ, Vandemark GJ (2007) Assessment of genetic diversity among nondormant and semidormant alfalfa populations using sequence-related amplified polymorphisms. Crop Sci 47(6):2274–2284

Bertrand A, Castonguay Y, Bourassa J (2014) A whole-plant screening test to identify genotypes with superior freezing tolerance. In: Hincha DK, Zuther E (eds) Plant cold acclimation methods and protocols. Methods in molecular biology, vol 1166. Humana Press, pp 35–42

Brummer EC, Cassler MD (2014) Cool-season forages In: Smith S, Diers B, Specht J, Carver B (eds) Genetic gain in majot U.S. field crops. CSSA Spec. Publ. 33. ASA, CSSA, and SSSA. Madison, WI

Cunningham S, Gana J, Volenec J, Teuber L (2001) Winter hardiness, root physiology and gene expression in successive fall dormancy selection from Mesilla and CUF101 alfalfa. Crop Sci 41:1091–1098

Dalmannsdottir S, Rapacz M, Jørgensen M, Østrem L, Larsen A, Rødven R, Rognli OA (2016) Temperature before cold acclimation affects cold tolerance and photoacclimation in timothy (*Phleum pratense* L.), Perennial Ryegrass (*Lolium perenne* L.) and Red Clover (*Trifolium pratense* L.). J Agron Crop Sci 202(4):320–330

Li X, Alarcón-Zúñiga B, Kang J, Nadeem Tahir MH, Jiang Q, Wei Y, Reyno R, Robins JG, Brummer EC (2015) Mapping fall dormancy and winter injury in tetraploid alfalfa. Crop Sci 55:1995–2011

Schwab PM, Barnes DK, Sheaffer CC (1996) The relationship between field winter injury and fall growth score for 251 alfalfa cultivars. Crop Sci 36(2):418–426

Stout DO, Hall JW (1989) Fall growth and winter survival of alfalfa in interior British Columbia. Can J Plant Sci 69:491–499

Teuber LR, Taggard KL, Gibbs LK, McCaslin MH, Peterson MA, Barnes DK (1998) Fall Dormancy. Standard tests to characterize alfalfa cultivars. North American Alfalfa Improvement Conference. http://naaic.org/stdtests/Dormancy2.html. Accessed 20 Feb 2017

Application of Multivariate Analysis for Genotype Evaluation in Soybean

V. Perić[(✉)], M. Srebrić, A. Nikolić, and S. Mladenović-Drinić

Maize Research Institute "Zemun Polje", Belgrade, Serbia
vesnaperic@yahoo.com

Abstract. Multivariate analysis is a frequently used approach in breeding studies. Applied to multiple trait data, Principal Component Analysis (PCA) gives an opportunity to graphically display the relationships between important agronomic traits, evaluate the varieties on the basis of multiple traits, identify valuable breeding sources and recommend possible selection strategies. This study was conducted to evaluate agronomic performance of 31 early varieties (maturity group 0) from the soybean collection of the Maize Research Institute "Zemun Polje" (Belgrade, Serbia). PCA analysis was applied to identify the best performing genotypes considering multiple traits (seed yield, major yield components and parameters of technological quality of grain), and to determine the level of trait interdependence. Genotypes were tested in a randomized complete block design with three replications during two years (2011 and 2012) at two locations in Serbia (Zemun Polje and Pančevo). An analysis of variance (mixed model) exhibited significant effects of genotype (G), environment (E), and genotype × environment (G × E) for most of the traits. According to PCA biplot, seed yield per plant was in positive correlation with number of seeds and pods per plant, indicating that those yield components could be effective selection criteria in breeding for seed yield. Comparison of early-maturing genotypes based on multiple traits revealed two varieties with above-average seed yield per plant and four varieties with high oil and above average protein content, which could be used as potential parents in breeding for important agronomic traits in future breeding programs.

Keywords: Agronomic traits · Correlations · PCA

1 Introduction

Favourable composition of soybean grain makes this crop major protein and oilseed source in many regions of the world. In Serbia, total soybean harvested area was 154249 ha (FAOSTAT 2014), mainly occupied by varieties that belong to 0, I and II maturity group. Although the seed yield is one of the most important breeding goals in soybean breeding programs, released varieties have to meet the requirements of processing industry regarding chemical composition and technological quality of grain. Development of high yielding and good grain quality cultivars is a demanding task due to the negative correlation between the protein and oil content, on one hand (Chung et al. 2003; Bellalui et al. 2009), and negative correlation between proteins and seed yield on the other (Li and Burton 2002; Taški-Ajduković et al. 2010). Thus,

© Springer International Publishing AG, part of Springer Nature 2018
G. Brazauskas et al. (Eds.): *Breeding Grasses and Protein Crops in the Era of Genomics*, pp. 215–219, 2018.
https://doi.org/10.1007/978-3-319-89578-9_39

identification of superior genotypes considering multiple traits is the primary goal of many breeding programs (Yan and Rajcan 2002). PCA is a useful method for evaluation of cultivars based on multiple traits, identification of those that are superior in desired traits (Mohammadi and Amri 2011), and visualization of genetic correlation among traits, as well (Yan and Rajcan 2002). The objective of this study was to identify superior genotypes as potential parents in future breeding cycles, examine the correlations between traits and determine effective selection criteria by applying the method of PCA.

2 Materials and Methods

The plant material for this study comprised 31 soybean accessions of the maturity group 0 (early maturing genotypes) from the soybean collection of the Maize Research Institute "Zemun Polje", originated from different parts of the world. The trials were carried out during two growing seasons (2011 and 2012), at two locations in Serbia (Zemun Polje and Pancevo) and arranged according to randomized block design with 3 replications and two rows per genotype. The experimental plot size was 5 m^2. Standard agricultural practice was applied. The soil type is slightly calcareous chernozem. During the both research years, particularly in 2012., severe drought occurred, with total rainfalls significantly lower and average temperatures much higher than multiyear average. The samples consisted of 30 plants per genotype collected at the R8 stage. A total of eight agronomically important traits were analyzed: *PH* – plant height, *NN* – node number, *PN* – pod number, SN – seed number, *TSW* – 1000 seed weight, *SYP* – seed yield per plant, *PROT* – protein content, *OIL* – oil content. Seed samples were analyzed for protein and oil content (expressed as a percentage on a dry matter basis) using grain analyzer Infraneo, Chopin Technologies ®. The data recorded for each agronomic trait were subjected to analysis of variance (mixed model). PCA was performed in order to determine trait interrelationships, to compare genotypes by multiple traits and identify candidates that could be used as parents. Relationships of traits revealed by biplot were compared to Pearson's correlation coefficients.

3 Results and Discussion

Analysis of variance of the mixed model showed high significant differences for all the sources of variation (Table 1). Predominant influence of environment was determined for all examined traits; the smaller part is attributed to genotype's effects, while the genotype × environment interaction contributed least to total variation.

First two axes (PC1 and PC2) on PC biplot accounted for 59.0% of the variation of the standardized data (Fig. 1). Along the PC1, genotypes were mostly allocated on the basis of *PN*, *SN*, *SYP* and *PROT*, while PC2 was mainly related to *PH*, *NN*, *TSW* and *OIL*. Majority of genotypes were intermediate for a number of traits, distributing close to the biplot origin. Biplot revealed group of genotypes (Turska 2, Apache, Atlas, Lambert) characterized by high oil and above average protein content (21% and 39%, respectively; data not shown). Highlighted varieties may be of importance for the

Table 1. Mean squares from analysis of variance for eight agronomic traits in soybean

Source of variation	Trait							
	PH	NN	PN	SN	TSW	SYP	PROT	OIL
G	48.8**	34.1**	20.8**	17.0**	65.4**	22.6*	51.9**	46.9**
E	831.7**	349.4**	420.4**	318.3**	752.4**	629.1**	239.4**	116.9**
G × E	8.7**	9.0**	5.5**	3.9**	14.9**	4.8**	9.2**	6.6**

PH – plant height, *NN* – node number, *PN* – pod number, *SN* – seed number, *TSW* – 1000 seed weight, *SYP* – seed yield per plant, *PROT* – protein content, *OIL* – oil content, *G* – genotype, *E* – environment, *G* × *E* – genotype × environment interaction
*P < 0.05; ** P < 0.01

simultaneous improvement of both traits, which are in soybean negatively correlated. Variety K-1 had the highest *PROT* and maximum *TSW*, but due to a lower *SYP* compared to the mean of the group, this variety has a potential for breeding for the grain size and protein content but not for the seed yield.

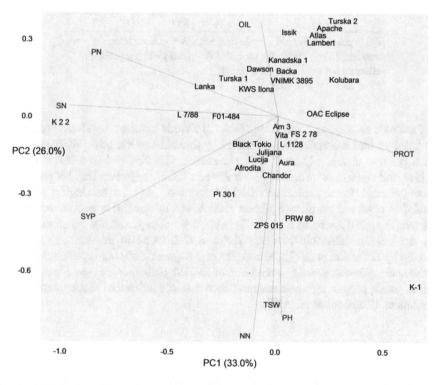

Fig. 1. PCA biplot of 31 soybean genotypes for 8 agronomic traits (*PH* – plant height, *NN* – node number *PN* – pod number, *SN* – seed number, *TSW* – 1000 seed weight, *SYP* – seed yield per plant, *PROT* – protein content, *OIL* – oil content)

Genotype K 2 2 had the largest *PN* and *SN*, while for *TSW* was intermediate, achieving the highest *SYP* in the group. Cultivar Lanka produced larger *PN*, *SN* and *TSW* as compared to the group average, and achieved the second ranked *SYP*. Those genotypes are a useful source for breeding for seed yield through a parallel selection of multiple components, in soybean usually negatively correlated.

The angles of trait vectors on the biplot suggest that the seed yield per plant showed positive correlation with *SN*, *PN* and *TSW*, negative correlation with *PROT*, and a weak negative correlation with *OIL*.

Table 2. Pearson's correlation coefficients between the traits

	NN	PN	SN	TSW	SYP	PROT	OIL
PH	0.72**	−0.23	−0.06	0.20	0.04	−0.06	−0.05
NN		−0.06	0.13	0.29	0.25	0.16	−0.11
PN			0.81**	−0.32	0.51**	−0.21	0.17
SN				−0.15	0.78**	−0.43*	0.07
TSW					0.49**	0.08	−0.29
SYP						−0.37*	−0.09
PROT							−0.57**

PH – plant height, *NN* – node number, *PN* – pod number, *SN* – seed number, *TSW* – 1000 seed weight, *SYP* – seed yield per plant, *PROT* – protein content, *OIL* – oil content
*P < 0.05; ** P < 0.01

Pearson's correlation coefficients (Table 2) were in agreement with the correlations displayed by biplot. Similar to the findings by Sudaric and Vrataric (2002), seed yield had a positive and highly significant correlation with *SN*, and to a lesser extent a positive and significant correlation with *PN* and *TSW*, suggesting that *SN* present the one of the most effective selection criteria. A moderate, negative and highly significant correlation observed for oil and protein content was in agreement with several studies (Popovic et al. 2013; Chung et al. 2003). Although not expected, the negative, low and non-significant relationship between *SYP* and *OIL* found in this research was also reported by Eskandari et al. (2013) and Yan and Rajcan (2002), and could be explained by the heterogeneous genetic background of studied genotypes and specific interaction of each genotype with the environment, which possibly influence the consistency of the correlations (Bellaloui et al. 2009).

4 Conclusion

Comparison of early-maturing soybean varieties based on multiple traits highlighted those that could be used as potential parents in breeding for agronomically important traits. Among the observed yield components, seed and pod number per plant were determined as the most reliable selection criteria.

Acknowledgments. This study is a part of a scientific project TR 31068 supported by the Ministry of Education, Science and Technological Development of the Republic of Serbia.

References

Bellaloui N, Smith JR, Ray JD, Gillen AM (2009) Effect of maturity on seed composition in the early soybean production system as measured on near-isogenic soybean lines. Crop Sci 49:608–620

Chung J, Babka HL, Graef GL, Staswick PE, Lee DJ, Cregan PB, Shoemaker RC, Specht JE (2003) The seed protein, oil, and yield QTL on soybean linkage group I. Crop Sci 43:1053–1067

Eskandari M, Cober ER, Rajcan I (2013) Genetic control of soybean seed oil: II. QTL and genes that increase oil concentration without decreasing protein or with increased seed yield. Theor Appl Genet 126(6):1677–1687

FAOSTAT (2014) Food and agriculture organization. http://faostat.fao.org/

Li H, Burton JW (2002) Selecting increased seed density to increase indirectly soybean seed protein concentration. Crop Sci 42:393–398

Mohamadi R, Amri A (2011) Graphyc analysis of trait relations and genotype evaluation in durum wheat. J Crop Improv 25:680–696

Popović V, Miladinović J, Vidić M, Tatić M, Sikora V, Ikanović J, Dozet G (2013) Productive and quality characteristics of soybean in agroecological conditions of Sombor, Serbia. Field Vegetable Crops Res 50(2):67–74

Sudarić A, Vratarić M (2002) Variability and interrelationship of grain quantity and quality characteristics in soybean. Die Bodenkultur 53(3):137–142

Taski-Ajdukovic K, Djordjevic V, Vidic M, Vujakovic M (2010) Subunit composition of seed storage proteins in high protein soybean genotypes. Pesquisa Agropecuária Brasileira 45(7):721–729

Yan W, Rajcan I (2002) Biplot analysis of test sites and trait relations of soybean in Ontario. Crop Sci 42:11–20

Yan W, Fregeau-Reid J (2008) Breeding line selection based on multiple traits. Crop Sci 48:417–423

Polymorphism of Soybean Cultivars and Breeding Lines Revealed by Marker *Satt100* Associated with the *E7* Locus

D. Zharikova[1], S. Ivanyuk[2], G. Chebotar[1], O. Korniychuk[2], and S. Chebotar[1(✉)]

[1] Odessa I.I. Mechnikov National University, Odessa, Ukraine
s.v.chebotar@onu.edu.ua
[2] Institute of Feeds and Agriculture of Podillia of NAAS, Vinnitsa, Ukraine

Abstract. Flowering time is critical for successful seed production of plants, and with time to maturity it determines geographic adaptation, seed quality and yield. In contrast to most cereals, soybean cultivars (*Glycine max*) are confined to comparatively narrow ranges of geographical latitudes. For every 100–150 km of changes latitude (corresponding to 1°), a new cultivar has to be developed. The aim of our work was to detect alleles of the microsatellite locus *Satt100*, which is tightly linked with the locus E_7 determining the time to flowering in soybean cultivars and breeding lines and to analyze any associations of the alleles with the time to flowering. Twenty five cultivars and breeding lines from the breeding program of the Institute of Feeds and Agriculture of Podillia of NAAS (IFAP; Ukraine) that cultivated in the Vinnitsa region of Ukraine were analyzed. Five alleles of *Satt100* locus have been detected in the plant materials studied. Only two lines KR-II №3035 and PSV-I №4001 have allele 'A' of *Satt100* that corresponds to dominant *E7* allele.

The alleles that have been detected at microsatellite locus *Satt100* for soybean breeding and parental lines of IFAP and their associations with different time to flowering is the first step for precise observations of the agronomical traits of the lines and estimations of their productivity with the purpose to find genotypes that have best adaptability to conditions of the Ukraine.

Keywords: *Glycine max* · Soybean · Photoperiod response · Flowering time
Microsatellite markers · E_7 gene

1 Introduction

Soybean (*Glycine max* (L.) Merr.) is one of the world's economically most important legume crops. Due to the unique combination of high content of protein, oil and carbohydrates, soybean is gaining popularity every year in the Ukraine where soybean growth on the acreage that increased from 73 thousand hectares in 2001 to 1.4 million hectares in 2013, and its production increased from 74 thousand to 2.8 million tons (Catalog of Soybean varieties 2014).

The photoperiod response is an important criterion that determines the latitudinal adaptation of soybean varieties. Maturity loci, by controlling time to flowering and

© Springer International Publishing AG, part of Springer Nature 2018
G. Brazauskas et al. (Eds.): *Breeding Grasses and Protein Crops in the Era of Genomics*, pp. 220–225, 2018.
https://doi.org/10.1007/978-3-319-89578-9_40

maturity, make soybean cultivars able to adapt to various regions with different day lengths (Dissanayaka et al. 2016). The locus E_7 is one of the eight previously reported loci that are involved in control of the time to flowering and maturity of soybean in temperate zones (Cober and Voldeng 2001). The E_7 locus shows a response to light quality, suggesting that it may be related to phytochrome (Monlar et al. 2003). The E7 allele had inhibitory effects for both browning and cracking (Benitez et al. 2004). Molnar et al. (2003) had mapped E_7 locus with the microsatellite markers Satt100, Satt319 and Satt460 on LG C2 (chromosome 6) on the region, that is spanning a 22.2 cM. According to Dissanayaka et al. (2016) E_7 was also linked with locus E_1 at approximate distance of 6.1 cM.

The dominant allele E7 is expressed in the induction phase of flowering and causes a delay to time to flowering on 4-5 days, that significantly increases the productivity (21% yield increase) (Rosenzweig et al. 2008). For Russia and Belarus conditions early flowering combined with an extended reproductive stage is preferable for soybean (Miroshnichenko 2005; Rosenzweig et al. 2003). There were no investigations of genes that affect time to flowering (involved in mechanism of photoperiodic sensitivity of plants) with molecular markers in Ukrainian soybean lines and varieties previously. Inclusion of this allele to the breeding material in Ukraine could be beneficial (Zharikova et al. 2016). Thus, optimization soybean breeding and development of new varieties for the conditions of Ukraine and regions with similar climate and investigation of flowering time inheritance in soybean germplasm are important. Preferred breeding strategy is to shift flowering time to earlier period, while the total vegetation period remains constant.

The aim of this work was detection of alleles of the microsatellite locus Satt100, which is tightly linked to the locus E_7, in the soybean cultivars and breeding lines of the IFAP breeding program and analysis of associations between time to flowering and different alleles of Satt100.

2 Materials and Methods

We have analyzed 5, 7 and 7 breeding lines F_{8-10} from three crosses, respectively – 'Oksana' x 'Labrador', 'Mapple Belle' x 'Sreska 72', 'Line 103' x 'Korada' that have been developed in IFAP. Totally twenty five cultivars and breeding lines, including six parental lines, that involved in the breeding program of the IFAP have been analyzed. 'Harosoy' isoline 'OT 89-5' and cultivar 'Vilana', which both have dominant allele E7 were used as controls for molecular analysis. The cultivars and breeding lines were grown up in field in three replicates using plots 0.9 m² and 1.8 m², with inter row spaces 45 cm in 2014–2016 years in Vinnitsa region of Ukraine (49° N latitude). Standard density was 50-55 plants per m². The soybean lines were sown in the field when the optimal temperature of soil at the depth 10 cm was 10–12 °C, and in 2014 it was 3–4/05; in 2015-2016 – 26-30/04. The conditions of vegetation were diverse, but generally all three years of research were characterized by elevated air temperatures and a lack of rainfall especially at the time of flowering and further vegetation period. Phenological observations: registration of flowering start (R1) was carried out when 10% of plants start to flowering according to "The methods of state testing of agricultural crops variety" (2001). The differences between the R1 was estimated by LSD of the corresponding significance level for the factor "Line".

DNA was isolated by using the NeoPrep100 DNA kit («Neogen Laboratory» , Ukraine) from seeds and/or green leaves. PCR was carried out as recommended by Molnar et al. (2003) for *Satt100* (forward primer – 5'-ACCTCATTTTGGCATAAA-3' and reverse primer – 5'-TTGGAAAACAAGTAATAATAACA-3'). PCR products were separated on 7% polyacrylamide gels and stained with $AgNO_3$ according to Promega's protocol (1999). Size of fragments of amplifications was determined by comparing with the marker of molecular weight pUC19/Msp I. The marker contains the following 13 discrete fragments (in base pairs): 501, 489, 404, 331, 242, 190, 147, 111, 110, 67, 34, 34, 26.

The mathematical treatment of the results was performed with computer program Statistica v.7. The influence of tested factors was determined by analysis of variance (ANOVA). For the quantitative estimation of the influence of the factor - "Allele" on the variance of the investigated characteristics, the parameter "influence of the factor" $\left(p_A^{in}\right)$ as recommended (Rokitskiy 1973) was used.

3   Results and Discussion

For the investigated material five alleles of microsatellite *Satt100* have been revealed, the four of which have been named according to Rosenzweig et al. (2008) 'A', 'B', 'D', 'E' and the fifth 'F' – was firstly tested in our experiment. Results of PCR-analysis of the locus *Satt100* and number of days to flowering time are presented in Table 1.

Table 1. The time to flowering of soybean cultivars and breeding lines in Vinnitsa region and alleles of locus *Satt100* associated with E_7 locus

Parents/lines created by hybridization	R1, days (Mean ± Std dev.)	Allele of *Satt100* locus	Allele of E_7 gene
♀ 'OKSANA'(p.f.)	45,6 ± 7,5 [a]	A	E_7
♂ 'LABRADOR'(p.f.)	26,5 ± 3,5 [b]	F	?
SR-II №2365	32,0 ± 3,6 [b]	F	?
SR-II №2367	30,3 ± 2,3 [b]	D	e_7
SR-II №2369	29,7 ± 3,1 [b]	F	?
KR-II №3035	36,0 ± 7,2 [ab]	A/F	E7?
PSV-I №4001	38,7 ± 3,8 [a]	A	E7
♀ 'MAPLE BELLE'(p.f.)	29,5 ± 0,7 [b]	E	e_7
♂ 'SRESKA 72' (p.f.)	35,0 ± 1,4 [ab]	E	e_7
SR-II №2372	38,7 ± 3,8 [a]	E	e_7
SR-II №2375	41,7 ± 4,6 [a]	E	e_7
SR-II №2377	43,5 ± 4,9 [a]	E	e_7
KR-I №3014	38,7 ± 7,2 [a]	E	e_7
KR-II №3045	35,3 ± 4,0 [ab]	E	e_7

(continued)

Table 1. (*continued*)

Parents/lines created by hybridization	R1, days (Mean ± Std dev.)	Allele of *Satt100* locus	Allele of E_7 gene
PSV-I №4005	38,7 ± 6,7 ^{a}	E	e_7
PSV-II №4017	37,7 ± 7,4 ^{ab}	E	e_7
♀ 'LINE 103' (p.f.)	35,6 ± 2,1 ^{ab}	E	e_7
♂ 'KORADA' (p.f.)	30,3 ± 2,3 ^{b}	B	e_7
SR-I №2220	37,0 ± 4,6 ^{ab}	E	e_7
SR-I №2226	38,0 ± 7,0 ^{ab}	E	e_7
SR-I №2234	46,0 ± 5,6 ^{a}	E	e_7
SR-I №2245	43,3 ± 6,7 ^{a}	E	e_7
SR-I №2250	43,3 ± 4,5 ^{a}	E	e_7
SR-I №2265	38,3 ± 6,8 ^{a}	E	e_7
SR-I №2359	39,3 ± 3,5 ^{a}	E	e_7
LSD$_{LINE0,05}$	8,61	–	–
LSD$_{LINE0,01}$	11,74	–	–
LSD$_{ALLELE0,01}$	12,05	–	–
P, %	2,01	–	–
p_A^{in}, %	64,0	–	–

Note. Letters (a, b) for the different similarity groups are given according to LSD$_{LINE0,01}$

Compare to control lines – 'Harosoy' isoline 'OT89-5' and cultivar 'Vilana' that are carriers of dominant allele *E7* and allele 'A' at *Satt100* as was shown Rosenzweig et al. (2008) we have revealed that parental line 'Oksana' has the PCR-fragment with the same size as controls, thus it has allele 'A'. For the parental line 'Labrador' we have revealed 147 bp amplification fragment, which we designed as fragment 'F'. Among five derivate lines from crossing 'Oksana' x 'Labrador', one – PSV-I №4001 was also characterized by fragment 'A' and line KR-II №3035 was heterogenic – some plants of this line have 'A' fragment while other plants have fragment specific to parental line 'Labrador'. Two derivate lines from the same crossing – SR-II №2365 and SR-II №2369 has also PCR-fragment as 'Labrador' ('F'); and line SR-II №2367 has PCR-fragment with the size 145 bp, that Rosenzweig et al. (2008) was assigned like 'D'. We assumed that for line SR-II №2367 crosspollination was occurred. The differences in R1 between two parental lines 'Oksana' and 'Labrador' was 19 days in average for 3 years that was significant (P > 0,01). Also significant differences (P > 0,01) for R1 were observed between genotypes of derivates from this crossing that carriers of allele *E7* ('A' fragment according to PCR-analysis of *Satt*100).

From the crossing 'Mapple Belle' x 'Sreska 72' both parental lines were characterized by PCR-products 149 bp, that we corresponded to 'E' fragment at the *Satt 100* according to nomenclature proposed by Rosenzweig et al. (2008). These authors assigned as a recessive allele e_7 all alleles of *Satt100* that differed from fragment 'A' specific to genotypes carriers of dominant allele E_7 such as 'Harosoy' isoline 'OT 89-5' and 'Vilana'. Thus in our experiment for crossing 'Mapple Belle' x 'Sreska 72' both parental lines have recessive e_7 gene. Among derivate lines from this crossing we also didn't revealed genetic polymorphism at the locus *Satt 100*, so the lines have allele e_7. However there was 5,5 days (not significant) difference in average R1 between parental lines. The time to flowering ranged from 35 days to 44 days for derivate lines with an average - of 39.4 days, although parental lines demonstrated tendency to earlier flowering (at 29,5 'Mapple Belle' and 35 day 'Sreska 72'). Seven lines were derived from the cross 'Line 103' x 'Korada'. Cultivar 'Korada' was earlier characterized by Rosenzweig et al. (2008) as a genotype with recessive allele e_7, that have been characterized by 'B' fragment of PCR-product of *Satt 100*. In our experiment, an amplification fragment of 149 bp have been revealed for 'Line 103', this fragment was assigned as 'E'. Thus, parents were polymorphic at locus *Satt 100*, but all lines selected from this crossing were not polymorphic and have 'E' fragment. The time to flowering for these lines ranged from 37 to 46 days, with an average of 40.7 days, and three lines SR-I №2234, SR-I №2245, SR-I №2250 started to flower significantly (LSD 0,01) later than 'Korada'. Despite the fact that parents were polymorphic at *Satt100* and have average 5,3 days of differences at R1 this difference was not significant according to LSD 0,05. That can be explained by the concordance of both fragments 'B' and 'E' to recessive allele e_7 gene.

4 Conclusion

Breeding lines created in IFAP and cultivars involved in breeding program were tested with the help of molecular marker *Satt 100*. We have found 5 alleles of *Satt 100* locus in the investigated material.

Cultivars 'Oksana' and 'Labrador' (parental lines of first crossing) have significant differences in time to flowering (P > 0,01). Lines from crossing 'Oksana' x 'Labrador' did not differ (P > 0,01) from each other and from 'Labrador' except line PSV-I №4001. The line PSV-I №4001 flowered significantly (P > 0,01) later than 'Labrador'. Such difference in FT could be explained by the effect of dominant allele of E_7 gene. We have not detected significant differences in to R1 between PSV-I №4001 and 'Oksana', but there was observed tendency to about 7 days earlier flowering than 'Oksana'.

There were no differences both in time to flowering (R1) and alleles at *Satt100* locus between 'Mapple Belle' and 'Sreska 72'. Thus it is interesting, that two lines SR-II №2375 and SR-II №2377, which are derivates from this crossing ('Mapple Belle' x 'Sreska 72'), flowered significantly later (P > 0,01) than 'Mapple Belle' and did not differ from 'Sreska 72'. Perhaps such difference could be explained by presence of other E genes in the genotype of parental lines.

In crossing 'Line 103' x 'Korada' was no significant differences in time to flowering between parental lines. We have revealed different alleles at *Satt100* locus, but both alleles had been associated with recessive allele e_7. Derivate lines SR-I №2234, SR-I №2245, SR-I №2250 flowered significantly later than parental line 'Korada'. All derivate lines had not polymorphism at *Satt100* and had the same allele like 'Line 103'.

The next step of our investigations will be to detect other alleles of *E* genes and analyze associations between genetic polymorphism at loci, that determine time to flowering, and yield in conditions of Vinnitsa region of Ukraine.

Acknowledgments. This study was funded by Ministry of Education and Science of Ukraine as research project № 569 «Polymorphism of wheat and soybean photoperiod sensitivity loci and plant development dependence on their allelic composition according to PCR analysis».

References

Benitez ER, Funatsuki H, Kaneko Y, Matsuzawa Y, Bang SW, Takahashi R (2004) Soybean maturity gene effects on seed coat pigmentation and cracking in response to low temperatures. Crop Sci. 44:2038–2042

Catalog of Soybean varieties (2014) http://fri.vin.ua/download_materials/catalog_soya_2014.pdf. pdf

Cober ER, Voldeng HD (2001) A new soybean maturity and photoperiod-sensitivity locus linked to E1 and T. Crop Sci 41:698–701

Dissanayaka A, Tito OR, Di S, Yan F, Githiri SM, Rodas FR, Abe J, Takahashi R (2016) Quantitative trait locus mapping of soybean maturity gene E5. Breed Sci 66:407–415. https://doi.org/10.1270/jsbbs.15160

Miroshnichenko MV (2005) Changes in agronomic traits caused by breeding in soybean. PhD thesis, Krasnodar (in Russian)

Molnar SJ, Rai S, Charette M, Cober ER (2003) Simple sequence repeat (SSR) markers linked to *E1*, *E3*, *E4*, and *E7* maturity genes in soybean. Genome 46:1024–1036

Promega Technical Manual (1999). Gene Print. STR Systems. Printed in USA. Revised. vol 7, 52 p

Rokitskii PF (1973) Biologicheskaya statistika (Biological Statistics). Vysheish Shk, Minsk

Rosenzweig VE, Aksyonova EA, Milash SB, Goloenko DV, Davydenko OG (2008) Prospects of exploiting of photoperiod sensitivity gene *E7* in early soybean breeding and revealing of its sources with SSR-markers. Soybean Genetics Newsletter, vol 35. http://www.soygenetics.org/articleFiles/

Rosenzweig VE, Goloenko DV, Davydenko OG, Shablinskaya OV (2003) Breeding strategies for early soybeans in Belarus. Plant Breed 122:456–458

Zharikova D, Ivanyuk S, Voytkova V, Chebotar S, Korniychuk O (2016) Polymorphism cultivars and lines of soybean Podillia determined by molecular marker *Satt100* associated with E_7 gene photoperiodic sensitivity. National scientific and practical internet conference on "Improving the efficiency of agriculture in climate change mitigation", held at the Institute irrigated agriculture NAAS (Ukraine), Kherson, 9 December 2016

Genetic Diversity Assessment of Two *Medicago sativa* Genes: CAD and WXP1

C. Gréard[1,2(✉)], P. Barre[1], S. Flajoulot[2], S. Santoni[3], and B. Julier[1(✉)]

[1] P3F, INRA, 86600 Lusignan, France
{camille.greard,bernadette.julier}@inra.fr
[2] Jouffray-Drillaud, INRA, 86600 Lusignan, France
[3] AGAP, INRA, 34060 Montpellier, France

Abstract. Alfalfa (*Medicago sativa* L.) is a major perennial forage legume crop with numerous nutritional and environmental benefits. This allogamous and autotetraploid species whose varieties are synthetic populations, has a highly polymorphic genome and a short linkage disequilibrium. Allele mining strategy on targeted candidate genes is an option to select valuable parents for breeding. In this paper, we assessed allelic diversity of CAD and WXP1 genes, involved in lignin biosynthesis and drought tolerance respectively, in a set of 384 individuals. For CAD which had two splice forms, 30 and 31 variants (out of 1077 and 906 base pairs of the coding sequences, respectively) have been observed including 37 and 52% of non-synonymous mutations. More variants were observed in WXP1 sequence, with 157 observed in WXP1 coding sequence (coding sequence of 1116 base pairs), including 60% non-synonymous mutations. Among the non-synonymous mutations, some probably affected protein function. Both genes were under purifying selection, especially CAD with dN/dS rate of 0.05 and 0.12 for both splice forms, against a rate of 0.26 for WXP1. Difference in variant proportion is probably explained by differential selective pressure that may be induced by contrasted expression levels. Indeed, CAD is highly and continually expressed whereas WXP1 is induced in some specific conditions. Further studies assessing the impact of variants on phenotype will help to conclude on the allele mining strategy in alfalfa breeding.

Keywords: Alfalfa · Lucerne · Selection · Mutation · Lignin biosynthesis
Drought tolerance

1 Introduction

Alfalfa (*Medicago sativa* L.) is a perennial autotetraploid species (2n = 4x = 32). This high nutritional value forage crop is able to fix atmospheric nitrogen (N) and also present numerous environmental benefits such as energy saving by its contribution to soil N fertilization, carbon and greenhouse gas retention, soil protection against erosion, biodiversity preservation (Huyghe et al. 2014).

However, because of the short linkage disequilibrium and the high genetic variability of alfalfa in populations and varieties (Herrmann et al. 2010), molecular biology tools such as QTL detection, association genetics and genomic selection are challenging to apply on breeding programs (Li et al. 2014).

© Springer International Publishing AG, part of Springer Nature 2018
G. Brazauskas et al. (Eds.): *Breeding Grasses and Protein Crops in the Era of Genomics*, pp. 226–230, 2018.
https://doi.org/10.1007/978-3-319-89578-9_41

Digestibility and drought resistance are two important traits in alfalfa breeding. For each trait, one gene with major effect on phenotype has been identified: CAD for digestibility and WXP1 for drought tolerance.

CAD catalyses the reduction of cinnamaldehydes to cinnamyl alcohols, that is the last step of monolignol biosynthesis. A down regulation of CAD induces a lower syringyl/guaiacyl ratio, that tends to increase alfalfa *in situ* dry matter digestibility (Baucher et al. 1998). The CAD gene studied here is an orthologue of *Medtr1g107425* identified in *Medicago truncatula*. This gene is the closest relative of CAD2 of *Eucalyptus gunnii* identified as the CAD gene with the strongest effect on lignin biosynthesis (Poke et al. 2003). It also corresponds to the *Medicago sativa* CAD sequence used for a transgenic plant study (Baucher et al. 1999). Two splice forms exist for this gene in *Medicago truncatula*, here called CAD-1 and CAD-2. CAD-1, with 4 exons, corresponds to a protein of 358 AA, identified as *Medtr1g107425.1*, and CAD-2 with 5 exons corresponds to a protein of 301 AA, *Medtr1g107425.2*.

The gene WXP1 is involved in wax production. Its transcription is induced by abscisic acid, cold and drought treatments. WXP1 overexpression in transgenic alfalfa enhances drought tolerance with increased accumulation of cuticular waxes (Zhang et al. 2005). Transformed *Arabidopsis thaliana* plants expressing *Medicago truncatula* WXP1 also have an increased freezing tolerance (Zhang et al. 2007). The gene studied here is an orthologue of *Medtr5g062700*, which was used for transgenic *M. truncatula* studies (Zhang et al. 2007).

In this paper, we will evaluate diversity of CAD and WXP1. Both genes are involved in traits of interest for the breeder and diversity information can be a clue to select parents for breeding.

2 Material and Methods

2.1 Plant Material and Genotyping

A set of 384 individuals of alfalfa was established, consisting of 48 cultivated varieties with about 9 individual per variety. Primers surrounding each target gene were defined in conserved parts of *M. sativa*'s genome, defined by comparison of a provisional genome assembly (J. Gouzy et al., unpublished) and sequences of two *M. sativa* genotypes. After DNA extraction, target genes of each individual were amplified by a long-range PCR using a high fidelity TAQ polymerase (Platinium, Invitrogen). PCR products were randomly sheared and used to build libraries suitable for Miseq sequencing. The libraries were sequenced at INRA of Montpellier with *Illumina* technology (MiSeq 2 × 250 bases).

CLC workbench (CLC genomics) was used to analyse sequences. Firstly, reads were trimmed on quality and size in order to keep only reads with a Phred score upper than 30 and to discard reads below 60 base pairs. Because of the bad quality of the provisional genome assembly of alfalfa, we built a new alfalfa consensus sequence for each gene. These consensus sequences were obtained by aligning a subset of 16 750 random reads (mean coverage of 100) on the parts of provisional genome assembly sequence that corresponded to the target genes.

Reads of each gene of each genotype were aligned on these alfalfa consensus sequences and the "variants" (SNP and InDel) were called. Only variants with a frequency above 10% and a read depth above 30 per position were retained. Our alfalfa consensus sequences were used to calculate dN/dS ratio by comparison with the *M. truncatula* coding DNA sequences (CDS). dN/dS corresponds to the ratio between non-synonymous (dN) and synonymous (dS) substitution rates. A dN/dS value smaller than 1 indicates that purifying selection tends to purge non-synonymous mutations at the evolutionary level. On the contrary a value above 1 occurs when some non-synonymous mutations are under positive selection.

3   Results and Discussion

The CDS of the two splice forms CAD-1 and CAD-2, 1077 and 906 base pairs long, respectively, contained a similar number of variants (Table 1). However, in the introns, CAD-1 contained more variants than CAD-2. This difference could be explained by the fact that CAD-1 had one more intron than CAD-2. WXP1 had no intron and was 1116 base pairs long. This gene had 4.5 times more variants than CAD. In regard to the CDS length, 12% of the sequence was concerned by variations, which was about four times more than for CAD.

Table 1. Variants observed in CDS and introns of the two splice forms of CAD (CAD-1 and CAD-2) and WXP1

	CDS				Introns			
Gene	Number of variants	Number of SNP	Number of InDel	% of variants in CDS sequence*	Number of variants	Number of SNP	Number of InDel	% of variants in intron sequence*
CAD-1	30	29	1	2.78	166	73	93	14.34
CAD-2	31	28	3	3.42	127	54	73	15.58
WXP1	137	117	20	12.28	–	–	–	–

*% of CDS sequence = number of variants in CDS/CDS length in base pairs; *% of intron sequence = number of variants in introns/total intron length in base pairs

The proportions of InDel over the total number of variants in CDS were low for both splice forms of CAD with 3.3% for CAD-1 and 9.7% for CAD-2. The number of InDel in WXP1 CDS was twice higher than in CAD-2 with 21.9% of variants being InDel. However, it was even higher in intron sequences with 55.7% for CAD-1 and 57.5% for CAD-2.

Among the variations that occurred in the CDS, some of them were coding for the same amino acid as the reference (synonymous variants), due to the degeneracy of the genetic code. For CAD-1, the proportion of synonymous variants over the total number of variants (63%) was nearly twice of the proportion of variants affecting the protein (non-synonymous variants). The proportion of synonymous substitutions for CAD-2 (52%) was lower than for CAD-1. No variant affecting the reading frame was found in CAD gene.

By contrast, 60% of WXP1 variants were non-synonymous. Among them, 5 InDels affected WXP1's reading frame, each carried by one individual and the other 76 non-synonymous variants did not impact reading frame.

We analyzed the relative importance of selection at the evolutionary level on each gene by calculating dN/dS ratio between *M. sativa* consensus CDS and *M. truncatula* reference CDS. Indeed dN/dS values were 0.057 and 0.119 for CAD-1 and CAD-2, respectively, suggesting that CAD gene is under a strong purifying selection, either natural or through breeding. With a dN/dS value of 0.264, WXP1 was also under influence of selection against non-synonymous mutations.

The lower proportion of variants in coding sequence and the higher synonymous mutation rate in CAD than in WXP1 suggested that CAD protein function is more affected by mutations than WXP1, so the mutations are more easily wiped out. This could be explained by the high expression level of CAD. Indeed it has been shown that the evolutionary rate (dN) is negatively correlated to the expression level (Drummond et al. 2005; Pál et al. 2001). High protein conservation of highly expressed genes can be a selective way to act against protein-misfolding and to preserve protein-protein interaction efficiency (Drummond et al. 2005; Omer et al. 2017; Yang et al. 2012). Thus, higher rate of non-synonymous variants and higher dN/dS ratio of CAD-2 splice form sequence compared to CAD-1 suggests that CAD-1 splice form could be more expressed than CAD-2.

The rate of non-synonymous substitution in WXP1 was particularly high and this gene probably accumulated more mutations than CAD because of irregular expression level. Indeed, WXP1 expression is induced by cold or drought stress so as long as the plant is grown under favorable conditions, WXP1 does not express and thus, no selection of any kind occurs (Zhang et al. 2005). A gene in repressed state with no apparent benefit for the organism is more likely to be subject to deletions, such as those we saw in WXP1 (Omer et al. 2017).

4   Conclusion

Both genes are under purifying selection either natural or induced by breeding process, but WXP1 is more likely to accumulate mutations than CAD. Non-synonymous mutations were observes in both genes. Crosses could be performed to assess their impact on their phenotype. Breeding strategy could then be set up in order to cumulate favorable alleles. This strategy would be particularly interesting to create low lignin varieties with mutated CAD genes.

Acknowledgments. PhD thesis of C. Gréard is supported by ANRT (French Ministry of Higher Education and Research), CIFRE Convention n°2015/1447.

References

Baucher M, Monties B, Montagu MV, Boerjan W (1998) Biosynthesis and genetic engineering of lignin. Crit Rev Plant Sci 17:125–197

Baucher M, Bernard-Vailhé MA, Chabbert B, Besle J-M, Opsomer C, Montagu MV, Botterman J (1999) Down-regulation of cinnamyl alcohol dehydrogenase in transgenic alfalfa (*Medicago sativa* L.) and the effect on lignin composition and digestibility. Plant Mol Biol 39:437–447

Drummond DA, Bloom JD, Adami C, Wilke CO, Arnold FH (2005) Why highly expressed proteins evolve slowly. Proc Natl Acad Sci USA 102:14338–14343

Herrmann D, Barre P, Santoni S, Julier B (2010) Association of a CONSTANS-LIKE gene to flowering and height in autotetraploid alfalfa. Theor Appl Genet 121:865–876

Huyghe C, Vliegher AD, van Gils B, Peeters A (2014) Grasslands and herbivore production in Europe and effects of common policies. Editions Quae

Li X, Han Y, Wei Y, Acharya A, Farmer AD, Ho J, Monteros MJ, Brummer EC (2014) Development of an alfalfa SNP array and its use to evaluate patterns of population structure and linkage disequilibrium. PLoS ONE 9:e84329

Omer S, Harlow TJ, Gogarten JP (2017) Does sequence conservation provide evidence for biological function? Trends Microbiol 25:11–18

Pál C, Papp B, Hurst LD (2001) Highly expressed genes in yeast evolve slowly. Genetics 158:927–931

Poke FS, Vaillancourt RE, Elliott RC, Reid JB (2003) Sequence variation in two lignin biosynthesis genes, cinnamoyl CoA reductase (CCR) and cinnamyl alcohol dehydrogenase 2 (CAD2). Mol Breed 12:107–118

Yang J-R, Liao B-Y, Zhuang S-M, Zhang J (2012) Protein misinteraction avoidance causes highly expressed proteins to evolve slowly. Proc Natl Acad Sci 109:5158–5159

Zhang J-Y, Broeckling CD, Blancaflor EB, Sledge MK, Sumner LW, Wang Z-Y (2005) Overexpression of WXP1, a putative Medicago truncatula AP2 domain-containing transcription factor gene, increases cuticular wax accumulation and enhances drought tolerance in transgenic alfalfa (*Medicago sativa*). Plant J 42:689–707

Zhang J-Y, Broeckling CD, Sumner LW, Wang Z-Y (2007) Heterologous expression of two *Medicago truncatula* putative ERF transcription factor genes, WXP1 and WXP2, in arabidopsis led to increased leaf wax accumulation and improved drought tolerance, but differential response in freezing tolerance. Plant Mol Biol 64:265–278

Testing for Freezing Tolerance
in Perennial Ryegrass

A. Aleliūnas$^{(\boxtimes)}$ and G. Brazauskas

Lithuanian Research Centre for Agriculture and Forestry, Institute of Agriculture,
Instituto a. 1, 58344 Akademija, Kėdainiai District, Lithuania
andrius.aleliunas@lammc.lt

Abstract. Perennial ryegrass (*Lolium perenne* L.) is a highly valued species both for agriculture and amenity purposes. However, it lacks winter hardiness for cultivation at certain regions. As winter hardiness is a very complex trait which is affected both by numerous environmental conditions and genetic composition of a plant, freezing tolerance is an important component of winter hardiness. The freezing tolerance at the population scale might be determined at selected fixed-temperature points or by the LT_{50} evaluation. Here the LT_{50} value evaluation has a potential to reveal substantially more information about this trait, as it involves assessment of survival at a series of freezing temperatures. Herein we present a high-throughput freezing tolerance evaluation method for perennial ryegrass populations.

Keywords: *Lolium perenne* · LT_{50} · Frost tolerance

1 Introduction

Perennial ryegrass is an outcrossing species and due to its strict self-incompatibility mechanism, it exists as highly genetically heterogeneous populations. Therefore, perennial ryegrass breeders might have a special interest in the performance of particular populations opposed to individual genotypes. Population screening requires a suitable amount of plants for the genetic diversity which lies within to be represented. Therefore, there is a need for a simple and high-throughput method to accomplish this task. Moreover, there is still a lack of unified methodology of artificial freezing screening and it makes the comparison of the results obtained at different laboratories impossible as minor differences in freezing experiment conditions might lead to inconsistency of the results.

2 Materials and Methods

2.1 Plant Material and Growth Conditions

LT_{50} values for the populations were determined by the temperature of the substrate. The substrate of 50/50 (vol) perlite/vermiculite mix was used for the experiment and the plants were established from seed in 4×7 horticulture cell packs with 100 ml

G. Brazauskas et al. (Eds.): *Breeding Grasses and Protein Crops in the Era of Genomics*, pp. 231–236, 2018.
https://doi.org/10.1007/978-3-319-89578-9_42

capacity per cell. Twenty seeds per population were sown in each cell and in total 3 replications per population were used for each target temperature in this experiment. Plants were grown in a flood-drain system in the greenhouse for 4 weeks.

2.2 Cold Hardening Conditions

Two days prior cold hardening plants were moved to phytotron PlantMaster (CLF Plant Climatics GmbH, Germany) set at 10 °C, 200 μmol m^{-2} s^{-1} PAR, 12/12 h photoperiod and 80% relative air humidity. Afterwards, the plants were cold hardened for 14 days at 2 °C, 200 μmol m^{-2} s^{-1} PAR, 12/12 h photoperiod and 80% relative air humidity.

2.3 Freezing Test Under Controlled Conditions

The determination of LT$_{50}$ values was performed following a modified protocol proposed by Skinner and Garland-Campbell (2014). Six target temperatures of −4 °C, −6 °C, −8 °C, −10 °C, −12 °C and −14 °C were chosen for LT$_{50}$ calculations. Freezing tests were conducted in a walk-in freezing chamber PE 2412UY-LX (Angelantoni Industrie S. p.A, Italy) in the dark. In order to avoid desiccation induced by moving cold air streams inside the walk-in freezing chamber, an open-top polystyrene box with a size (1.7 × 1.6 × 0.3 m.) to fit 18 horticulture cell packs was constructed. After loading of the samples the top of the box was covered by agro fabric. The temperature in the freezing chamber was gradually lowered from 2 °C to −4 °C at the rate of 2 °C h^{-1} and then maintained for 76 h. Then the temperature was lowered at the rate of 2 °C h^{-1} until target temperature has been reached. Samples were held at the target temperature for 24 h and then the temperature was raised to 6 °C at a rate of 2 °C h^{-1} and held for 24 h allowing the substrate to defrost.

2.4 Plant Recovery After Freezing and Survival Evaluation

Survival of the plants was evaluated after 21 days of recovery in the greenhouse. Plants demonstrating any growth were scored as survived.

2.5 Substrate Temperature Logging During the Freezing Test

The temperature of the substrate was measured by the means of thermocouple probes in each horticulture cell pack. Thermocouple probes were inserted in the substrate at a depth of approx. 2.5 cm. A multi-purpose data loggers KD7 TYPE (Lumel S.A., Poland) with twelve standard measuring channels for each unit was used to log substrate and air temperature data every 2 min.

2.6 Calculation of LT$_{50}$ Values

The dose-response data was subsequently used for LT$_{50}$ calculation. A custom R script was written to facilitate the LT$_{50}$ calculation in panels consisting of a larger number of populations. Generalized linear model using a binomial distribution to fit a curve was employed and MASS package (Venables and Ripley 2002) for R (R Core Team 2015) was used to calculate fractional dosage value of 0.5 and to determine an LT_{50} value for each population.

2.7 Statistical Analyses

All statistical analyses were performed using R (R Core Team 2015).

3 Results and Discussion

Freezing tolerance evaluation in perennial ryegrass still remains labour-intensive and costly as it requires substantial man hour input as well as electricity cost is relatively high, especially when a series of freezing tests must be performed to collect the survival data for LT_{50} determination.

One must take caution when performing freezing tests as a number of, at first glance minor factors can considerably affect the results. Moreover, even under controlled-freezing conditions, it is difficult to achieve a uniform temperature distribution across the experiment panel. This was also observed during our experiments despite a relatively small experimental freezing area of 1.7×1.6 m.

During the initial freezing test stage, there was steady and almost uniform decrease in substrate temperature. However, at the moment of ice formation, a sudden and a sharp rise in substrate temperature could be observed. Subsequent exothermal heat dissipation was slow and the temperature inconsistency across the freezing panel was substantial. Therefore, it is necessary to track the temperature by the temperature probes inserted into the substrate and a data logger, in order to introduce corrections and to minimize the effect of a temperature gradient. Subsequent corrections according to the actual substrate temperature improve the accuracy of the freezing test (Fig. 1). While testing a large perennial ryegrass population panel for freezing tolerance we found that LT_{50} ranged from −9.81 to −5.37 °C. Despite a large number of populations tested, our results can not be directly compared to similar studies. LT_{50} values obtained in our study were substantially higher as compared to those determined by Sugiyama (1998) where the most freeze-susceptible accession had an LT_{50} value of −12.7 °C; in our study the most freezing-tolerant population had an LT_{50} of only −9.81 °C. The results we obtained during our experiment were much more similar to those of Ebdon and Gagne (2002). Interestingly, substantial differences in obtained LT_{50} values were also observed between Ebdon and Gagne (2002) and Sugyama (1998) studies, despite it shared a similar sample preparation methodology as the plant samples were washed free of soil, packed to poly-freezer bags and subjected to freezing test. Again, there was no overlap of the LT_{50} values as the lowest of cold-acclimated perennial ryegrass plants obtained in Ebdon and Gagne (2002) study was −12.6 °C and, despite a similar workflow of sample preparation, LT_{50} values obtained by Sugyama (1998) were significantly lower with the highest of −12.7 °C. In this case, we can not overrule the possibility that the tested plant material had different levels of freezing tolerance. However, most likely, the differences in freezing test conditions led to the LT_{50} value differences between those two experiments.

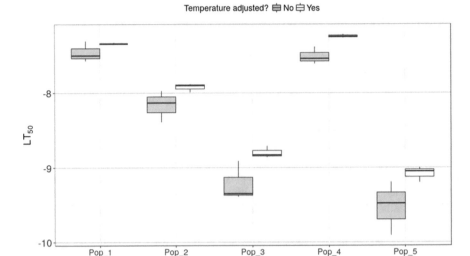

Fig. 1. LT$_{50}$ values derived from on air temperature inside the freezing chamber versus substrate temperature adjusted by the means of thermocouple probes.

Therefore, the freezing tests must be performed under the same settings as possible, unless the researcher has an interest of comparing the effect of these parameters on the freezing tolerance trait. There is a must to standardize pot sizes (if the freezing test is done on whole plants grown in pots/trays as it affects temperature dynamics of the substrate), growing substrate, growth, cold-acclimation and, especially, freezing test conditions and duration. Special attention must be taken when transferring plants after the freezing test for regrowth, as the temperature regime might affect the survivability considerably (**data not shown**).

In addition, freezing tolerance of a plant might differ at distinct development stages. Lemežys (1997) tested the freezing tolerance of several perennial ryegrass cultivars under controlled conditions. It was found that tetraploid cultivars were generally more freezing susceptible as compared to their diploid relatives, especially at the stage of young seedling. However, as the plant development advanced, the differences became less apparent.

As we tested diploid and tetraploid perennial ryegrass populations for freezing tolerance, we found the superiority of diploid population in freezing tolerance as compared to the tetraploid ones (p-value < 0.0001) (Fig. 2). However, it is still unclear, what physiological factors led to the observed differences in freezing tolerance between these groups. One possible explanation for higher levels of freezing tolerance could be the smaller cell size in diploids, as the degree of the cell contraction under freezing conditions is likely to be more reduced as compared to the tetraploid cell (Limin and Fowler, 1994).

Nevertheless, the freezing test results are difficult to compare even in those settings when the freezing tests are performed at the same laboratory as the freezing tests are separated by time. Those observed differences might arise due to the systematic errors.

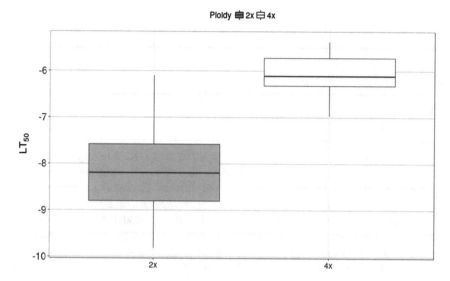

Fig. 2. Freezing tolerance of diploid and tetraploid perennial ryegrass populations.

However, the sensitivity of the freezing method is quite high. In our case, two separate perennial ryegrass population panels were tested for freezing tolerance. Paired t-test revealed the existence of statistically significant differences (p-value = 0.0001) in LT_{50} values obtained from two independent experiments. The mean of the LT_{50} differences between those two panels was $-1.04°C$ (Fig. 3). However, in this case, Pearson

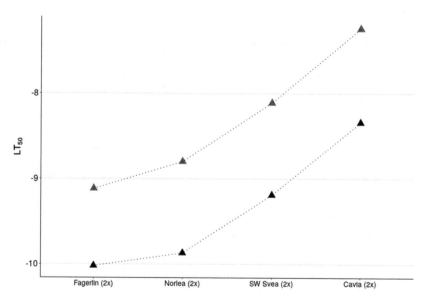

Fig. 3. LT_{50} values of the same populations obtained during two independent experiments performed approx. one year apart. Despite the significant differences in obtained LT_{50} values, the correlation in this case was found to be extremely high.

correlation coefficient between LT_{50} values obtained during these two experiments was extremely high (r = 0.99, p-value = 0.003). One possible way to standardize the freezing test results could be the inclusion of several standard populations in every freezing test and adjust the results according to the survival of standard populations (Skinner and Garland-Campbell, 2014).

4 Conclusion

Artificial freezing tests offer an excellent opportunity to test the freezing tolerance in a large number of perennial ryegrass populations in a relatively short period. However, the results are difficult to compare between the different laboratories. Moreover, they also might be inconsistent at the same laboratory settings as a number of minor factors might affect the outcome. Therefore, it is highly recommended to test the freezing tolerance in the populations of interest altogether, especially, if the test panel fits freezing chamber. Otherwise, the inclusion of standard populations in each experiment is recommended to enable the possibility to compare the results obtained in distinct experiments.

Acknowledgments. The study was carried out in the frame of the Development of Molecular Markers for Genomic Selection of Adaptation in Perennial Ryegrass project (MIP-064/2015) funded by the Research Council of Lithuania and the Private Public Partnership on Pre-breeding in Perennial Ryegrass project funded by the Nordic Council of Ministers.

References

Ebdon JS, Gagne RA, Manley RC (2002) Comparative cold tolerance in diverse turf quality genotypes of perennial ryegrass. Hortic Sci 37:826–830
Lemežys E (1997) Cold hardiness of diploid and induced tetraploid perennial ryegrass (Lolium perenne L.). In: Proceeding of international conference 'plant breeding: theories, achievements and problems', Dotnuva-Akademija, pp 130–135
Limin AE, Fowler DB (1994) Relationship between guard cell length and cold hardiness in wheat. Can J Plant Sci 74:59–62
R Core Team (2015) R: a language and environment for statistical computing. R Foundation for Statistical Computing, Vienna, Austria. https://www.R-project.org/
Skinner DZ, Garland-Campbell K (2014) Measuring freezing tolerance: survival and regrowth assays. Methods Mol Biol 1166:7–13. https://doi.org/10.1007/978-1-4939-0844-8_2
Sugiyama S (1998) Differentiation in competitive ability and cold tolerance between diploid and tetraploid cultivars in lolium perenne. Euphytica. https://doi.org/10.1023/A:1018322821118
Venables WN, Ripley BD (2002) Modern applied statistics with S, 4th edn. Springer, New York

The Multifunctional Role of Legumes in Vineyards and Orchards

T. Vymyslický[(⊠)], J. Lang, and D. Knotová

Agricultural Research, Ltd., Troubsko, Czech Republic
vymyslicky@vupt.cz

Abstract. In frame of the research project TA04020464 "Different ways of vineyard greening and management and their influence on reduction of soil erosion and quality of production", various species of legumes have been studied as cover-crop mixtures. The main aim of the research is to recommend the best mixtures for the use in vineyards and orchards in both organic and integrated crop production. Based on two years evaluation of five mixtures in six localities, it can be concluded that the best results were obtained by winter mixture, and perennial mixture for dry conditions, and annual mixture for fast greening. These three mixtures comply with the requirement of fast vegetation-cover establishment after sowing, and thus significantly reduce the erosion.

Keywords: Agrotechnology · Cover crops · Erosion · *Fabaceae*
Seed mixtures

1 Introduction

Greening of vineyards and orchards is used from many positive reasons, especially suppression of weed growth, increase of organic matter content and nutrient content in the soil, reduction of erosion, and improvement of conditions for mechanisation **movement**. Well established greening positively influences growth of grape and fruit trees. The aim of modern management in vineyards and orchards is transition from spontaneous greening to higher species diversity in the undergrowth, which better influences grape plants, fruit trees, and grapes and fruits. For the greening, wide spectrum of plant species from different plant families can be used (Hirschfelt 1998; Hartwig and Ammon 2002).

Leguminous plants have many positive ecological roles. In the vineyards and orchards they contribute to elimination of soil and wind erosion and serve as feed sources for pollinators, insects and vertebrates. Maybe their most important function is the ability to fix atmospheric nitrogen (Hartwig and Ammon 2002; Pardini et al. 2002). Since leguminous plants can supply other plants with nitrogen in acceptable form, they are essential parts of all organic and most integrated systems of production (Sulieman 2014). Grass–legume mixtures can provide more nitrogen than legume pure stands due to mutual stimulation of nitrogen uptake from symbiotic and non-symbiotic sources (Nyfeler et al. 2011).

© Springer International Publishing AG, part of Springer Nature 2018
G. Brazauskas et al. (Eds.): *Breeding Grasses and Protein Crops in the Era of Genomics*, pp. 237–242, 2018.
https://doi.org/10.1007/978-3-319-89578-9_43

That is why we focused in frame of the research project "Different ways of vineyard greening and management and their influence on reduction of soil erosion and quality of production" on the development and composition of new soil covering seed-mixtures with essential part of these mixtures being represented by leguminous plants for use in vineyards and orchards in both organic and integrated production systems. In this work we found that species-rich seed-mixtures, both for inter rows and undergrowth stripes for grape plants or fruit trees, supported biodiversity in the vineyards and orchards, and had positive effect on soil physical and chemical characteristics, growth and development of grape plants of fruit trees, yields and quality of grapes, fruits and vines.

2 Material and Methods

For our research on different ways of greening the vineyards and orchards the six following localities, situated in the south-eastern part of the Czech Republic, were selected: Hnízdo, Mikulov, Popice-Gotberg, Popice-Sonberk, Archlebov and Syrovín. All the localities are situated in the Pannonian part of the Czech Republic with warm and dry continental climate.

In March 2015 the following five soil-covering seed mixtures with important representation of leguminous plants were assembled:

Perennial species-rich mixture with *Anthyllis vulneraria, Festuca arundinacea, Festuca ovina, Festuca rubra, Hyssopus officinalis, Lotus corniculatus, Medicago lupulina, Onobrychis viciifolia, Origanum vulgare, Plantago lanceolata, Securigera varia, Trifolium pannonicum* and *Trifolium repens;*

Perennial mixture for dry conditions with *Festuca ovina, Festuca rubra, Medicago lupulina, Plantago lanceolata, Trifolium incarnatum* and *Trifolium repens;*

Annual mixture for fast greening with *Camelina sativa, Fagopyrum esculentum, Lolium multiflorum, Lotus ornithopodioides, Phacelia congesta, Phacelia tanacetifolia, Phalaris canariensis, Sinapis arvensis, Trifolium alexandrinum, Trifolium campestre* and *Trifolium resupinatum;*

Annual mixture for pollinators with *Calendula officinalis, Camelina sativa, Coriandrum sativum, Crambe abyssinica, Fagopyrum esculentum, Lolium multiflorum, Matricaria recutita, Phacelia congesta, Phacelia tanacetifolia, Phalaris canariensis, Satureja hortensis* and *Trifolium incarnatum;*

Winter mixture with *Anthyllis vulneraria, Daucus carota, Lolium multiflorum, Medicago lupulina, Trifolium incarnatum* and *Vicia pannonica.*

Component selection reflected not only component availability on the Czech seed-market, but also reasonable price for farmers. Two sowing rates were tested: – standard, according to mixture type 17–35 kg/ha); increased by 30%, according to mixture type 22–46 kg/ha). The sowing rate of each mixture was based on the ratio of individual components and their recommended sowing rates. Each mixture was sown on an area of 1200 m^2 in all localities. Ideal conditions consisted in three inter rows of 2 m width and 200 m length. The mixtures were sown in March 2015 and 2016.

Presence and cover of each sown species was evaluated in all localities after the emergence period in late spring (end of May) and in autumn (end of October) of 2015 and 2016. Based on presence - absence of sown species in three permanent plots of 10 × 2 msize for each mixture the average success was calculated for each locality.

3 Results and Discussion

Data of average share of functional groups within examined mixtures in Table 1 show that the most successful functional group were legumes (35.4% sown × 45.2% observed), grasses (42.4% sown × 33% observed) and herbs (22.2% sown × 21.8% observed).

Overall survey of the mixtures evaluation in experimental localities in the years 2015 and 2016 are presented in Table 2. The most successful soil covering seed mixture with important representation of leguminous plants in all tested localities was winter mixture (94% success), followed by perennial mixture for dry conditions (85%), annual mixture for fast greening (79%), annual mixture for pollinators (58%) and the worst results were observed for the perennial species rich mixture (55%). From the point of view of localities the most successful was Popice-Gotberg (84% success), followed by Mikulov (77%), Hnízdo (76%), Syrovín (73%), Archlebov (71%) and the worst results were observed in Popice-Sonberk (64%).

When we compare the results across localities, we found that the two closest localities, situated only two km far away, had the most different results with Popice-Gotberg as the best, and Popice-Sonberk as the worst. This fact could be explained by microclimate differences between Gotberg situated in a shallow valley with moist microclimate, having more suitable microclimatic conditions, than Sonberk, situated on dry and warm south-eastern slope.

Table 1. Share of functional groups within different mixtures

Mixture	Grasses sown	Grasses observed	Legumes sown	Legumes observed	Herbs sown	Herbs observed
Perennial species-rich mixture	40%	30%	58%	68%	2%	2%
Perennial mixture for dry conditions	60%	30%	39%	65%	1%	5%
Annual mixture for fast greening	40%	35%	30%	38%	30%	27%
Annual mixture for pollinators	22%	20%	5%	10%	73%	70%
Winter mixture	50%	50%	45%	45%	5%	5%
Average	42.4%	33.0%	35.4%	45.2%	22.2%	21.8%

Table 2. The results of success evaluation of five seed mixtures studied at six localities in the years 2015 and 2016.

Year 2015

Mixture	Perennial species rich		Perennial for dry conditions		Annual for fast greening		Annual for pollinators		Winter	
Sowing rate	Standard	Increased	Standard	Increased	Standard	Increased	Standard	Increased	Standard	Increased
Hnízdo	51%	57%	86%	87%	82%	81%	54%	57%	91%	93%
Mikulov	62%	56%	87%	83%	83%	86%	59%	54%	92%	96%
Popice-Gotberg	65%	67%	90%	96%	87%	84%	63%	71%	93%	98%
Popice-Sonberk	39%	37%	72%	73%	62%	74%	41%	43%	89%	87%
Archlebov	44%	49%	75%	81%	74%	85%	42%	49%	94%	95%
Syrovín	51%	56%	83%	87%	73%	81%	48%	51%	91%	96%

Year 2016

Mixture	Perennial species rich		Perennial for dry conditions		Annual for fast greening		Annual for pollinators		Winter	
Sowing rate	Standard	Increased	Standard	Increased	Standard	Increased	Standard	Increased	Standard	Increased
Hnízdo	54%	58%	86%	89%	84%	82%	63%	67%	97%	98%
Mikulov	68%	61%	91%	83%	78%	85%	64%	63%	92%	95%
Popice-Gotberg	76%	67%	92%	98%	95%	89%	67%	84%	97%	96%
Popice-Sonberk	42%	45%	75%	74%	61%	70%	47%	51%	84%	97%
Archlebov	54%	59%	85%	91%	77%	81%	42%	50%	95%	96%
Syrovín	51%	56%	83%	87%	73%	78%	58%	57%	93%	95%
Total 2015 + 2016	55%	56%	84%	86%	77%	81%	54%	58%	92%	95%

After comparing of two experimental seasons, we can conclude, that worse results were obtained in dry and hot season 2015, while in the mild year 2016 the results were better, particularly for the annual mixtures (Table 2). Nevertheless, the differences among years were statistically not significant.

Based on the first two years of trial evaluation we can conclude, that suitable mixtures should have significantly higher proportion of legumes – even higher than 50% (Carlsson and Huss-Danell 2003). Usual proportion of leguminous species in classical forage mixtures is significantly lower, in most cases from 10 to 20% (Sleugh et al. 2000). Dominance of grasses results in strong competition for water. The water could be in some cases depleted only by grasses. It results in strong deficit of water for grape plants and fruit trees.

For organic or integrated viniculture and orchard farming is much better higher proportion of legumes from the reason of fixation of atmospheric nitrogen and deeper rooting compared to grasses, enabling water infiltration into deeper soil layers (Bugg et al. 1996). Grape plants and fruit trees well supplied with nitrogen are healthier and more drought resistant.

The use of sufficient share of herbs and leguminous plants in the mixture is important in case of pollinators since – they provide feed sources for a long period and enable sustainable honey production in the vineyards and orchards (Decourtye et al. 2010).

4 Conclusion

The main aim of our research was to recommend the best seed mixtures for the use in vineyards and orchards in both organic and integrated production. Based on two years results of evaluation of five seed mixtures in six localities we can conclude that the best results were reached by winter mixture (94% success), followed by perennial mixture for dry conditions (85%) and annual mixture for fast greening (79%). These three mixtures fulfil the requirement of fast creation of vegetation cover after sowing and thus the reduction of erosion. That is why they could be recommended for the use in the vineyards and orchards of south Moravian region.

Acknowledgments. The results were obtained in the frame of the research project TA04020464 "Different ways of vineyard greening and management and their influence on reduction of soil erosion and quality of production", financed by the Technology Agency of the Czech Republic.

References

Bugg RL, Mc Gourty G, Sarrantonio M, Lanini WT, Bartolucci R (1996) Comparison of 32 cover crops in an organic vineyard on the north coast of California. Biol Agric Hortic 13:63–81
Carlsson G, Huss-Danell K (2003) Nitrogen fixation in perennial forage legumes in the field. Plant Soil 253(2):353–372
Decourtye A, Mader E, Desneux N (2010) Landscape enhancement of floral resources for honey bees in agro-ecosystems. Apidologie 41:264–277
Hartwig NL, Ammon HU (2002) Cover crops and living mulches. Weed Sci 50:688–699

Hirschfelt DJ (1998) Soil fertility and vine nutrition. In: Cover cropping in vineyards. a grower's handbook, pp 61–68

Nyfeler D, Huguenin-Elie O, Suter M, Frossard E, Lüscher A (2011) Grass–legume mixtures can yield more nitrogen than legume pure stands due to mutual stimulation of nitrogen uptake from symbiotic and non-symbiotic sources. Agr Ecosyst Environ 140(1):155–163

Pardini A, Faiello C, Longhi F, Mancuso S, Snowball R (2002) Cover crop species and their management in vineyards and olive groves. Adv Hortic Sci 16:225–234

Sulieman S, Tran L-SP (2014) Symbiotic nitrogen fixation in legume nodules: metabolism and regulatory mechanisms. Int J Mol Sci 15(11):19389–19393

Sleugh B, Moore KJ, George JR, Brummer EC (2000) Binary leg-ume–grass mixtures improve forage yield, quality, and seasonal distribution. Agron J 92:24–29

Hydroponic Evaluation of Growth and Nutrient Uptake in a *Lolium/Festuca* Introgression Series

D. Gasior[✉], J. Harper, A. Thomas, C. Evans, R. Mathews, D. Allen,
M. Humphreys, and I. Armstead

Institute of Biological, Environmental and Rural Sciences,
Aberystwyth University, Aberystwyth, UK
dkg@aber.ac.uk

Abstract. Introgression breeding represents a useful method for generating new germplasm resources which combine positive attributes from different species. Within forage grasses, the interfertility of *Lolium* and *Festuca* spp. represents a viable route for achieving this end. In order to evaluate the architecture of trait inheritance within this introgression context, a complete chromosome introgression series of *F. pratensis* (Fp; meadow fescue) in the *L. perenne* (Lp; perennial ryegrass) background, and derived progeny, were evaluated for shoot and root biomass and N and P uptake in a hydroponic experiment. The results indicated that the Fp parent and the monosomic introgressions relating to Fp chromosomes 3 and 4 showed enhanced shoot and root biomass. However, there was no clear relationship between these phenotypes of the Fp chromosome 3 and 4 monosomic introgressions and that of their derived, respective, progenies. Thus, the interaction between the Lp and Fp genomes in determining phenotype is likely to be complex.

Keywords: *Lolium perenne* · *Festuca pratensis* · Perennial ryegrass
Meadow fescue · Introgression · Biomass · Nutrient uptake · Hydroponics

1 Introduction

L. perenne (Lp; perennial ryegrass) and *F. pratensis* (Fp; meadow fescue) are widely grown forage grasses in N. Europe. Lp generally produces the highest quantity and quality of forage and is often the species-of-choice in more temperate regions. However, Fp becomes an increasingly important component of the sward in more northerly latitudes and continental climates, due to its greater tolerance to low-temperature and other abiotic stresses. As a consequence, it has become an aim of forage breeders to develop methods for combining the favourable attributes of these two species in 'festulolium' varieties, either using an amphiploid or an introgression breeding approach (Barnes et al. 2014; Humphreys et al. 1997; Kosmala et al. 2006; Macleod et al. 2013; Ostrem et al. 2013; Yamada et al. 2005) To this end, Harper et al. (2011) and King et al. (2013) reported the development of a monosomic introgression series for all 7 chromosomes of Fp in the Lp background (2n = 2× = 13 Lp + 1 Fp chromosome(s)) and derived progenies. Each of these progeny contained a single

G. Brazauskas et al. (Eds.): *Breeding Grasses and Protein Crops in the Era of Genomics*, pp. 243–248, 2018.
https://doi.org/10.1007/978-3-319-89578-9_44

haploid, introgressed sub-chromosome segment of Fp which, across the whole population covered the great majority of the Fp genome. In order to understand the architecture of trait inheritance in such a systematically developed Lp/Fp introgression series we have initiated a series of phenotype evaluations encompassing the entire set of parental genotypes, monosomic introgressions and derived progeny.

2 Material and Methods

2.1 Plant Material

The production of this Lp/Fp introgression series has been described in detail (Harper et al. 2011; King et al. 2013). Briefly, a tetraploid Lp genotype (Meltra) was crossed with a diploid Fp genotype to derive a triploid hybrid (LpLpFp). This was backcrossed to a diploid Lp genotype (Liprior) and 6 of the 7 monosomic introgressions (MSIs 1–6) of Fp in the Lp background were recovered as individual genotypes; for MSI 7 a further backcross to Liprior was required (MSI numbering, 1–7, identifies the single Fp chromosome introgressed into the Lp background according to the standard nomenclature). Each of these MSIs was then backcrossed to Liprior and 105 genotypes, each containing a single, introgressed, sub-chromosome segment of Fp, were identified. These 105 genotypes contained the majority of the Fp genome as evaluated cytologically and by molecular marker analysis. In addition, a Meltra x Liprior triploid hybrid was backcrossed onto Liprior to develop a control population (19 genotypes). For the analysis, the 124 genotypes were subdivided as 8 families according to their derivation from one of MSIs 1–7 (designated Family 1–7) or the control cross.

2.2 Hydroponic Conditions and Phenotype Evaluations

Six tillers of uniform size for each genotype were equilibrated for 28 days in 200 dm^3 of flowing nutrient solution (μM, NO_3^-, 250; K^+, 250; Ca_2^+, 344; SO_4^{2-}, 424; Mg^{2+}, 100; $H_2PO_4^-$, 50; Fe^{2+}, 5.4; with micronutrients as described by Clement et al. (1978)), replenished weekly. Air and solution temperatures were maintained at 20/15 $\pm$ 1 °C day/night and 20 °C, respectively with a 12 h photoperiod (650 $\pm$ 25 μmol m^{-2} s^{-1} PAR at canopy height). After 28 days the nutrient solution was replaced, as above, with the exception of NO_3^- at 50 μM, 1 atom % ^{15}N. Plants were grown for a further 28 days with nutrients added daily proportionate to an existing model of N availability (average, 255 mgN/d). Solution pH was maintained at 6.0 $\pm$ 0.1. Plants were harvested on day 28 and shoot and root dry weights (dwt) ascertained. Total N and ^{15}N enrichment were determined by IR-MS and total P was determined by ICP-OES with results expressed on a dry weight basis.

3   Results and Discussion

All of the overall comparisons for the studied traits analyzed using ANOVA indicated significant differences were present between the 8 populations ($p < 0.05$) and for all traits except shoot biomass, individual pairwise family differences were identified (Table 1). This indicated that, in spite of the narrow genetic base involved in the production of these families, considerable interfamily variation had been generated, particularly for shoot ^{15}N and P accumulation.

Table 1.   Tukey-Kramer significant pairwise comparisons between Families 1–7 and the control

	Shoot dwt*	Root dwt***	shoot[15]N***	root[15]N***	shootP***	rootP***
1 (8)	–	3, 4, 7, C	7, C	–	–	–
2 (13)	–	–	7	–	7, C	–
3 (23)	–	1	7, C	–	5	–
4 (16)	–	1, 5	7, C	–	7, C	7
5 (11)	–	4, 7	7, C	7	7, C	–
6 (13)	–	–	7	–	7, C	–
7 (21)	–	1, 5	1, 2, 3, 4, 5, 6	5	2, 4, 5, 6	4, C
C (19)	–	1	1, 3, 4, 5	–	2, 4, 5, 7	7

Number in brackets after the family name indicates the population size for that family. ANOVA for trait evaluation significant at $p < 0.05$* or $p < 0.001$***

Figure 1 illustrates the range of trait performances across the parental genotypes, MSIs and the 8 derived families. While differences between parental genotypes (with the exception of Meltra which grew poorly under hydroponic conditions), MSIs and their respective derived Families were not marked for N and P traits, it is apparent that for shoot and root biomass the Fp parental genotype noticeably outperformed the Lp parental genotypes, and that MSIs 3 and 4 also showed high scores for these traits – though this was not reflected in the mean Family 3 and 4 values. In order to evaluate as to whether individual genotype performance for shoot and root biomass could be attributed to the inheritance of a particular segment of Fp chromosome 3 and 4, mean scores for all genotypes within Families 3 and 4 containing the same region of intro-gressed Fp chromosome were compared (Fig. 2). While some individual genotypes showed trait scores approaching those of the Fp parent and MSIs 3 and 4, this analysis showed no consistent associations linking the presence of a particular Fp-derived chromosome segment with enhanced trait performance for shoot or root biomass were apparent. Thus, if the increased levels of shoot and root biomass seen in MSIs 3 and 4 were causally associated with the presence of the Fp chromosome, then the inconsis-tency with which the same trait performances manifested in derived progenies would indicate a more complex set of genome interactions than simple Fp segment introgression.

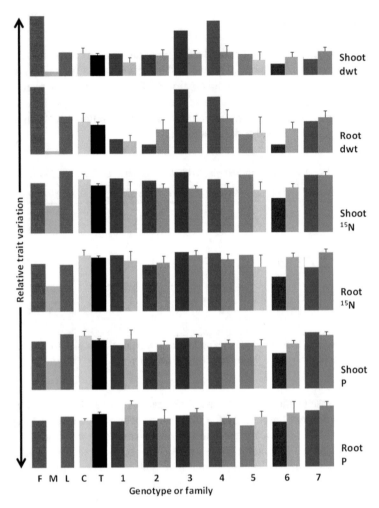

Fig. 1. Trait scores across the parental genotypes, MSIs 1–7 and Families 1–7. F - Fp parent, M - 4× Lp parent, L - 2x Lp parent, C - control population mean, T – Families 1–7 and control combined mean, 1 to 7 – MSIs 1–7 (left, darker colour) and respective, derived Family 1–7 means (right, lighter colour). Error bars represent 95% confidence intervals.

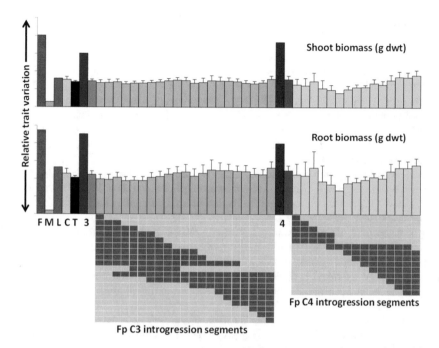

Fig. 2. Trait scores for shoot and root biomass illustrated for parental genotypes, MSIs 1–7 and derived Families 1–7. F - Fp parent, M - 4× Lp parent, L - 2× Lp parent, C - control family mean, T - Families 1–7 and control combined mean. Error bars, where n ≥ 3 - standard error; where n = 2, - range. *Fp chromosome (C) 3 and 4 introgression segments*: the individual genotypes within Families 3 and 4 are represented diagrammatically, in terms of the relative position of the introgressed Fp segment, as red and green squares below the bar chart. Each row is a single Family 3 or 4 genotype and each column indicates whether that genotype has a Fp introgression in that region (red) or not (green). Each light blue or orange bar is the trait mean and standard error/range for Family 3 or 4 genotypes with Fp introgressed segments (red) at that position.

4   Conclusion

The relationship between Fp chromosome introgression into the Lp genomic background and the quantitative trait biology of derived progeny is likely to be complex. While the transfer of traits with relatively simple genetic control may be possible through the manipulation of single, introgressed, chromosome segments, we will need to develop deeper understandings of the architecture of genome complementarities when dealing with more complex situations.

Acknowledgments. This study was funded by the UK Biotechnology and Biological Sciences Research Council (BBSRC) Crop Genetics, Genomics and Germplasm Institute Strategic Programme Grant (2012–2017).

References

Barnes BD, Kopecky D, Lukaszewski AJ, Baird JH (2014) evaluation of turf-type interspecific hybrids of meadow fescue with perennial ryegrass for improved stress tolerance. Crop Sci 54(1):355–365

Clement CR, Hopper MJ, Jones LHP (1978) The uptake of nitrate by Lolium perenne from flowing nutrient solution. I. Effect of NO3 - concentration. J Exp Bot 29:453–464

Humphreys M, Thomas HM, Harper J, Morgan G, James A, Ghamari-Zare A (1997) Dissecting drought- and cold-tolerance traits in the *Lolium-Festuca* complex by introgression mapping. New Phytol 137:55–60

Harper J, Armstead I, Thomas A, James C, Gasior D, Bisaga M, Roberts L, King I, King J (2011) Alien introgression in the grasses Lolium perenne (perennial ryegrass) and Festuca pratensis (meadow fescue): the development of seven monosomic substitution lines and their molecular and cytological characterization. Ann Bot 107(8):1313–1321

King J, Armstead I, Harper J, Ramsey L, Snape J, Waugh R, James C, Thomas A, Gasior D, Kelly R, Roberts L, Gustafson P, King I (2013) Exploitation of interspecific diversity for monocot crop improvement. Heredity 110(5):475–483

Kosmala A, Zwierzykowski Z, Gasior D, Rapacz M, Zwierzykowska E, Humphreys MW (2006) Gish/fish mapping of genes for freezing tolerance transferred from *Festuca* pratensis to *Lolium multiflorum*. Heredity 96:243

Macleod CJA, Humphreys MW, Whalley WR, Turner L, Binley A, Watts CW, Skot L, Joynes A, Hawkins S, King IP, O'Donovan S, Haygarth PM (2013) A novel grass hybrid to reduce flood generation in temperate regions. Sci Rep 3:1683

Ostrem L, Volden B, Larsen A (2013) Morphology, dry matter yield and phenological characters at different maturity stages of ×*Festulolium* compared with other grass species. Acta Agriculturae Scandinavica Section B-Soil and Plant Science 63(6):531–542

Yamada T, Forster JW, Humphreys MW, Takamizo T (2005) Genetics and molecular breeding in *Lolium/Festuca* grass species complex. Grassl Sci 51:89–106

Use of GBS for Lucerne Variety Distinction

B. Julier[1(✉)], P. Barre[1], P. Lambroni[1], S. Delaunay[1], F. Lafaillette[2],
M. Thomasset[3], and V. Gensollen[4]

[1] P3F, INRA, 86600 Lusignan, France
bernadette.julier@inra.fr
[2] GEVES, Domaine de l'Anjouère, 49370 Erdre-en-Anjou, France
[3] BioGEVES, Domaine du Magneraud, 17700 Surgères, France
[4] GEVES, 711 rue JF Breton, 34090 Montpellier, France

Abstract. Distinctness among lucerne varieties, evaluated in DUS tests, is a frequent obstacle to variety registration. Promising varieties may fail the distinction test despite an agronomic plus-value so the genetic progress is not delivered to farmers. The Genotyping By Sequencing (GBS) technology that enables to obtain a large number of markers on pools of individuals was used to evaluate the distinction among 20 lucerne varieties of the European Catalogue. A total of 40 000 polymorphic markers was obtained. All the varieties were statistically different from the others, and the structure was consistent with knowledge on the varieties. Perspectives to use GBS in DUS testing are given.

Keywords: *Medicago sativa* · Genetic diversity · Molecular markers
Genotyping By Sequencing

1 Introduction

Despite persistent genetic improvement of lucerne varieties, as showed by VCU tests, variety distinction is often difficult. For this reason, promising varieties may fail the distinction test and be never registered so the genetic progress is not delivered to farmers. Low throughput genotyping methods (AFLP or SSR) gave weak distinctness among lucerne varieties (Crochemore et al. 1996; Flajoulot et al. 2005). The Genotyping By Sequencing (GBS) technology that enables to obtain a large number of markers (Elshire et al. 2011) can be used on pools of individuals (Raineri et al. 2012) and has shown promising results to compare synthetic varieties of perennial ryegrass (Byrne et al. 2013) and lucerne (Annicchiarico et al. 2016). In the present UPOV regulations, molecular markers are not allowed for direct variety distinction, but they are used to structure the genetic diversity for a few species and to choose couples of varieties (applicant varieties and registered varieties) to grow side by side in the field during DUS testing (UPOV 2013). The present study was set up to evaluate the distinction among 20 lucerne varieties chosen in the European Catalogue.

© Springer International Publishing AG, part of Springer Nature 2018
G. Brazauskas et al. (Eds.): *Breeding Grasses and Protein Crops in the Era of Genomics*, pp. 249–253, 2018.
https://doi.org/10.1007/978-3-319-89578-9_45

2 Material and Methods

2.1 Plant Material

Twenty varieties (Table 1) obtained by different breeders have been chosen in the DUS reference catalogue to cover all the range of diversity for autumn dormancy, flower colour, disease (*Colletotrichum trifolii* and *Verticillium alboatrum*) and pest (*Ditylenchus dipsaci*) resistances. All varieties but one (Vernal) are registered in Europe.

2.2 Genotyping

Seeds were sown in a greenhouse and leaflets were sampled on four different bulks of 100 individuals for each variety. For each bulk, DNA extraction was carried out with a CTAB protocol. GBS librairies were built by using the restriction enzyme ApeK1 and a specific barcode for each bulk. Librairies were sequenced on an Illumina HiSeq 3000 at a the Genomic plateform GeT-PlaGe of INRA (INRA Toulouse, France) with a 150

Table 1. Description of the 20 alfalfa varieties

Variety	Flower colour[a]				Dormancy[b]	Resistance to[c]			Breeder	Registered in
	C1	C2	C3	C4		Va	Dd	Ct		
Arpège	1	8	2	1	4	7	7	4	Florimond Desprez	2004
Artémis	1	8	2	1	4	6	8	9	Barenbrug	2010
Barmed	1	8	2	2	7	5	5	8	Barenbrug	2002
Capri	1	7	2	1	4	7	7	4	Florimond Desprez	1995
Daphné	1	7	2	2	4	5	7	4	Florimond Desprez	1996
Dorine	1	7	3	1	6	3	5	3	Barenbrug	2001
Europe	1	8	2	1	4	5	3	na	Florimond Desprez	1961
Fado	1	8	2	1	4	6	8	6	Florimond Desprez	2012
Félicia	1	7	2	2	4	7	8	7	GIE Grass	2009
Franken Neu	1	5	1	6	4	3	4	na	Schmitz Ernst	1980
Galaxie	1	7	2	2	4	5	7	7	Semunion	2007
Greenmed	1	7	2	2	1	4	7	na	AgriObtentions	2011
Juurlu	3	2	1	1	1	na	na	na	Jogeva PBI	2002
Luzelle	1	6	2	3	3	3	6	na	INRA	1993
Meldor	1	6	2	2	6	4	5	4	INRA	1995
Midi	1	6	2	2	5	4	5	na	Tourneur	2002
Milky Max	1	7	2	2	4	7	8	6	GIE grass	2015
Orca	1	1	8	1	4	2	4	na	Carneau	1966
Verdor	1	7	2	2	6	5	5	3	Barenbrug	2011
Vernal	1	4	4	4	2	na	na	na	Wisconsin and Utah agricultural experiment stations	1953

[a] Intensity of flower color expression, from 1: 0–11% of plants to 9: 89–100% of plants with the indicated flower colour: C1: yellow, white or cream, C2: light purple, C3: dark violet, C4: variegated, [b] Autumn dormancy on the US scale ranging from 1 (very dormant) to 11 (non dormant); [c] Va: *Verticillium alboatrum*, Dd: *Ditylenchus dipsaci*, Ct: *Colletotrichum trifolii*; na: non available

base single-end protocol. A Linux pipeline was developed to obtain the allele frequencies. It contains the successive steps: (i) demultiplexing, (ii) trimming of the
sequences; (iii) alignment of the sequences on *M. truncatula* reference genome
sequence, version 4.0 with bwa, (iv) variant calling based on the GATK programme,
(v) calculation of the allelic frequency of each SNP in each bulk obtained by the
proportion of the variant allele to the total number of reads, (vi) merging of the files
obtained for all bulks into a single file.

Statistical analyses were based on DAPC (Discriminant Analysis of Principal
Components) with the procedure *dapc* of R software, calculation of F_{ST} between pairs
of varieties and test for significance of F_{ST} with the procedure *stamppFst* of R.

3 Results and Discussion

A total of 39 424 markers was obtained. The alleles with a frequency close to 0.25 or
0.75 in the populations were the most numerous. The comparison of the 4 bulks of 100
plants showed a good correlation within each variety (Fig. 1).

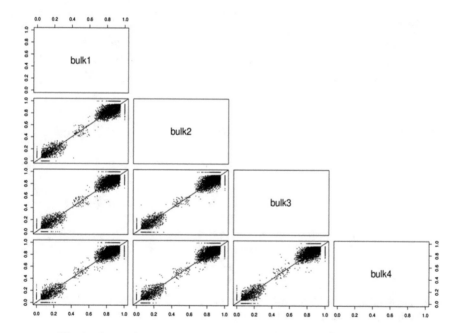

Fig. 1. Comparison of allelic frequency in the 4 bulks of variety Galaxie

With the DAPC analysis, five varieties were clearly separated from the 15 other
ones (Fig. 2): 1/Juurlu is a variety from Estonia, with a dormancy of 1 and a high
proportion of yellow flowers, 2/Vernal has a dormancy of 2, variegated flowers and
was bred for Northern USA and Canada, 3/Greenmed is a turf variety selected from

wild *M. sativa* populations, 4/Luzelle is a French variety adapted to grazing that contains one plant of the ssp *falcata* among the 13 parents of the initial polycross, and 5/Franken Neu is a German variety with a high proportion of variegated flowers. The other 15 varieties were all bred in France. When zooming on these 15 varieties, the 5 varieties adapted to the South (Midi, Meldor, Barmed, Dorine and Verdor) had the lowest values on the axis 2 than the 10 varieties adapted to the North (Artémis, Arpège, Fado, Capri, Europe, Félicia, Galaxie, Orca, Daphné, Milky Max) (Fig. 2).

The F_{ST} calculated between pairs of varieties ranged from 0.075 to 0.207 and were all highly significant ($P < 0.001$). This indicates that all varieties were different from the others.

On the contrary to other studies conducted with a limited number of markers (AFLP or SSR), we obtained a clear distinctness of varieties and a structure that is relevant to previous knowledge on the varieties. The high number of GBS markers obtained in the present study (close to 40 000) probably explained the better discrimination than in a study with 11 Italian landraces but only 2900 markers (Annicchiarico et al. 2016).

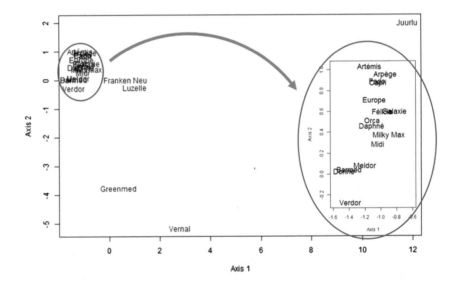

Fig. 2. Distribution of the 20 lucerne varieties on the first two axes of the DAPC. The small graph on the right is a zoom on the 15 surrounded varieties.

4   Conclusion

GBS markers, scored on bulks of individuals, proved efficient to differentiate lucerne varieties. This method could help in DUS testing. The reference collection of lucerne varieties could be genotyped with GBS markers. On this basis, taking example on the use of markers for maize distinction testing, groups of similar varieties could be established. When a new variety is submitted to registration, the first step could be to genotype it and to assign it to a group. In a second step, the phenotypic distinction tests could be carried out, within the group of similar varieties. Another strategy would be to base all distinction tests on molecular markers and to use phenotypic traits for variety description and VCU evaluation. These results can help the actors of variety registration to propose an evolution of the regulations, for better efficiency of the efforts devoted to lucerne variety distinctness.

Acknowledgments. This study was funded the French Ministry of Agriculture (project CAS-DAR Amediluze C2014-05). We thank the Genomic Plateform GeT-PlaGe of INRA in Toulouse, France, for sequencing.

References

Annicchiarico P, Nazzicari N, Ananta A, Carelli M, Brummer EC (2016) Assessment of cultivar distinctness in alfalfa: a comparison of genotyping-by-sequencing, simple-sequence repeat marker, and morphophysiological observations. Plant Genome 9:1–12

Byrne S, Czaban A, Studer B, Panitz F, Bendixen C, Asp T (2013) Genome wide allele frequency fingerprints (GWAFFs) of populations via genotyping by sequencing. PLoS One 8: e57438

Crochemore ML, Huyghe C, Kerlan MC, Durand F, Julier B (1996) Partitioning and distribution of RAPD variation in a set of populations of the *Medicago sativa* complex. Agronomie 16:421–432

Elshire RJ, Glaubitz JC, Sun Q, Poland JA, Kawamoto K, Buckler ES, Mitchell SE (2011) A robust, simple genotyping-by-sequencing (GBS) approach for high diversity species. PLoS One 6:e19379

Flajoulot S, Ronfort J, Baudouin P, Barre P, Huguet T, Huyghe C, Julier B (2005) Genetic diversity among alfalfa (Medicago sativa) cultivars coming from a single breeding program, using SSR markers. Theor Appl Genet 111:1420–1429

Raineri E, Ferretti L, Esteve-Codina A, Nevado B, Heath S, Perez-Enciso M (2012) SNP calling by sequencing pooled samples. BMC Bioinform. 13:239

UPOV (2013) Guidance of the use of biochemical and molecular markers in the examination of distinctness, uniformity and stability (DUS). http://www.upov.int/edocs/tgpdocs/en/tgp_15.pdf

Genome-Wide and Functional Gene Analysis of Perennial Ryegrass for Improved Growth Under Water Limiting Conditions (GrowGene)

K. Jonavičienė[(⊠)], G. Statkevičiūtė, and A. Aleliūnas

Institute of Agriculture, Lithuanian Research Centre for Agriculture and Forestry,
Akademija, Lithuania
kristina.jonaviciene@lammc.lt

Abstract. Breeding high yielding crops capable to adapt to future climate change and to meet the food needs of growing human population is of high priority. Perennial ryegrass, one of the most economically important forage grass species, performs poorly under water deficit conditions thus limits its growth and reduces the biomass production. The GrowGene project aims at identifying the genes, underlying biomass formation under water limiting conditions, by genotyping perennial ryegrass collection at a genome-wide scale and validating their function using emerging genomic and phenomic technologies. The knowledge gained could help to unlock the genetic mechanism, enabling the plants to cope with short term water limitation. Moreover, it could be used to improve biomass production under future climate conditions.

Keywords: Perennial ryegrass · Drought · Biomass formation
TILLING · Genotyping-by-sequencing · Phenotyping

1 Introduction

Due to an increase in the consumption of food, feed, fuel and to meet global food security needs for the rapidly growing human population, there is a necessity to breed for high yielding crops that can adapt to future climate change. Perennial ryegrass (*Lolium perenne* L.) is a species of paramount importance for modern pasture agriculture. However, it is more sensitive to abiotic stress compared to other cool season forage grass species. Perennial ryegrass exhibits poor performance under unfavourable environmental conditions such as drought and biomass production can be substantially reduced. Therefore, the changing climate poses a substantial threat to perennial ryegrass farming in continental regions. While there is significant variability in the tolerance of drought among different perennial ryegrass genotypes, the underlying genes and genetic mechanisms are still unknown.

In order to further understand the basic mechanisms leading to drought tolerance and to improve this trait by breeding, a perennial ryegrass panel, which has been characterised for growth parameters under water-limiting conditions using a high-throughput phenotyping platform, will be genotyped by sequencing. By exploiting Next-Generation Sequencing capabilities, this will allow us to build

© Springer International Publishing AG, part of Springer Nature 2018
G. Brazauskas et al. (Eds.): *Breeding Grasses and Protein Crops in the Era of Genomics*, pp. 254–258, 2018.
https://doi.org/10.1007/978-3-319-89578-9_46

high-density genome profiles that can be used for genome-wide association studies. Interesting candidate genes controlling growth traits under drought will be further evaluated for their functional role in the tolerance to changing environments process using TILLING. This reverse genetics technique allows to identify mutant alleles for any target gene, and to test the corresponding phenotypes. Since it does not fall under GMO regulations, the discovered valuable plant material can directly be translated to the field for crop improvement. The development of phenomic tools and the identification of genes affecting biomass production under drought conditions will address the challenges of breeding perennial ryegrass cultivars capable to produce stable biomass yield under water limiting conditions.

2   Concept

Future climate change in Europe, including Lithuania, will impact crop productivity. The physiological processes, controlling phenological traits of the plants, such as yield formation, drought stress tolerance, are adapted to local environments. The change of these conditions will demand new idiotypes of agricultural plants, which can produce stable yields under changing environment. Functional genome-wide analysis will enable to identify genes, underlying superior tolerance to water deficit thus will ensure stable yields and development of the molecular markers will facilitate breeding for enhanced drought tolerance of perennial plants. GrowGene project aims: to analyse perennial ryegrass genetic polymorphism on a genome-wide scale in order to identify the genes, underlying biomass formation under water limiting conditions and to validate their function using emerging genomic and phenomic technologies (Fig. 1).

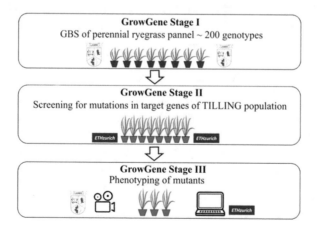

Fig. 1. Workflow of GrowGene project.

3 Approach

3.1 Plant Material and Phenotyping

To meet the predicted production demands for food, feed and fuel of the global population by 2050, crop production must double (Tilman et al. 2011). Extensive breeding and agronomic efforts over the past 50 years have tripled crop yields, but in recent decade the yields have been steadily declining and does not ensure world food security (Godfray et al. 2010). The environmental impacts, such as drought summers spells in temperate regions, are escalating crop demand thus it is of vital importance to develop crops with high and stable yields of good quality. This could be realized in future crops by combining a broad range of innovative breeding technologies and strategies (Tester and Langridge 2010).

Plant breeding is the process to genetically improve crops for desired traits by selection. It mainly consists of the generation of variability in the breeding nursery, selection and extensive testing of the selected material to either be developed into a new variety or to be introduced into the next breeding cycle (Manzanares et al. 2016). As the grasslands play a major role in milk and meat production (Wilkins and Humphrey 2003), breeding of perennial ryegrass (*Lolium perenne* L.) for increased biomass, which is the primary yield target, is very important. Given that biomass accumulation is largely determined by leaf growth, understanding the effect of water limitation on leaf growth could be used as a diagnostic tool to assess the plants' response to drought and to improve this trait through breeding (Tester and Langridge 2010). However, to non-invasively determine, when a plant perceives drought, is challenging. Therefore, a novel and high throughput phenotyping platform and statistical approach, developed by the project leader together with project partners from ETH Zurich (Jonavičienė et al. 2016a, 2016b) offers opportunity to assess growth in dynamic environment and is of value in dissecting complex data into quantitative parameters; which can be used as traits for further association analysis.

3.2 Candidate Gene Identification and Validation

Next generation sequencing (NGS) technologies rapidly developed over the last decade providing new opportunities to uncover the relationship between genotype and phenotype with greater resolution than ever before. NGS technologies are widely used for de novo sequencing, whole genome sequencing (WGS), whole genome re-sequencing (WGRS), genotyping by sequencing (GBS) as well as for transcriptome and epigenetic analysis (Varshney et al. 2009). The low cost of GBS per data point and its flexibility in terms of the proportion of the genome to be sequenced and the sequencing depth makes it an excellent tool to establish high-density genome-wide marker profiles (Beissinger et al. 2013). This tool detected thousands of SNPs in large populations of perennial ryegrass; it was also used for genetic diversity analysis and evolutionary studies (Lu et al. 2013) as well as genome-wide association studies (GWAS) in forage grasses. Most of the genetic studies in perennial ryegrass mainly focused on heading date and vernalization response, seed yield related traits, resistance to diseases (for review Kopecký and Studer 2014) and only few studies reported on candidate gene-based association analysis results on drought tolerance (Yu et al. 2013) and leaf growth under favourable environmental conditions (Auzanneau et al. 2011). This study will associate genome-wide polymorphisms and trait variation to identify genomic regions where genes governing leaf growth under water limiting conditions are located. To our knowledge, this has not been described so far.

The ability to associate a genome region to a certain trait stimulates the reverse genetics approaches to identify and characterize interesting phenotypes arising from manipulation of a particular gene (Jung et al. 2008). One of the method for functional gene analysis in grass species is Targeting Induced Local Lesions in Genomes (TILLING), known for successful application to identify knockout mutations and provide allelic mutations in target genes from pathogenic bacteria, animals, and plants. Several years TILLING populations were available for main crop plants, such as rice, wheat, sorghum and recently perennial ryegrass TILLING population has been developed (Manzanares et al. 2016). This method has been proven to decipher the function of the genes, particularly those that are unique to the species, e.g. perennial ryegrass.

In this way, the knowledge gained by application of next generation genomic and phenomic tools will help to understand the genome structure and function of perennial grasses and will provide tools for breeding stable yield formation under water limiting conditions of perennial ryegrass.

Acknowledgments. We thank prof. Bruno Studer and dr. Steven Yates for the agreement to collaborate. The research project is funded by the Research Council of Lithuania, grant No. S-MIP-17-24 (GrowGene)

References

Auzanneau J, Huyghe C, Escobar-Gutiérrez AJ, Julier B, Gastal F, Barre P (2011) Association study between the gibberellic acid insensitive gene and leaf length in a Lolium perenne L. synthetic variety. BMC Plant Biol 11:183

Beissinger TM, Hirsch CN, Sekhon RS, Foerster JM, Johnson JM, Muttoni G et al (2013) Marker density and read depth for genotyping populations using genotyping-by-sequencing. Genetics 193:1073–1081

Godfray HCJ, Beddington JR, Crute IR, Haddad L, Lawrence D, Muir JF et al (2010) Food security: the challenge of feeding 9 billion people. Science 327:812–818

Yu X, Bai G, Liu S, Luo N, Wang Y, Richmond DS et al (2013) Association of candidate genes with drought tolerance traits in diverse perennial ryegrass accessions. J Exp Bot 64:1537–1551

Jonavičienė K, Yates S, Nagelmüller S, Liebisch F, Kirchgessner N, Walter A et al (2016a) Modelling the growth of perennial ryegrass under water limiting conditions. In: Plant breeding: the art of bringing science to life, p 75

Jonavičienė K, Yates S, Nagelmüller S, Liebisch F, Kirchgessner N, Walter A et al (2016b) Real time growth analysis of perennial ryegrass. In: Breeding in a world of scarcity. Springer, Heidelberg, pp 105–110

Jung K-H, An G, Ronald PC (2008) Towards a better bowl of rice: assigning function to tens of thousands of rice genes. Nat Rev Genet 9:91–101

Kopecký D, Studer B (2014) Emerging technologies advancing forage and turf grass genomics. Biotechnol Adv 32:190–199

Lu F, Lipka AE, Glaubitz J, Elshire R, Cherney JH, Casler MD et al (2013) Switchgrass genomic diversity, ploidy, and evolution: novel insights from a network-based SNP discovery protocol. PLoS Genet 9:e1003215

Manzanares C, Yates S, Ruckle M, Nay M, Studer B (2016) TILLING in forage grasses for gene discovery and breeding improvement. N Biotechnol 33:594–603

Tester M, Langridge P (2010) Breeding technologies to increase crop production in a changing world. Science 327:818–822

Tilman D, Balzer C, Hill J, Befort BL (2011) Global food demand and the sustainable intensification of agriculture. Proc Natl Acad Sci USA 108:20260–20264

Varshney RK, Nayak SN, May GD, Jackson SA (2009) Next-generation sequencing technologies and their implications for crop genetics and breeding. Trends Biotechnol 27:522–530

Wilkins PW, Humphrey MO (2003) Progress in breeding perennial forage grasses for temperate agriculture. J Agric Sci 140:129–150

Genomic Selection for Biomass Yield of Perennial and Annual Legumes

P. Annicchiarico$^{(\boxtimes)}$, N. Nazzicari, L. Pecetti, and M. Romani

Research Centre for Fodder Crops and Dairy Productions (CREA-FLC),
Lodi, Italy
paolo.annicchiarico@crea.gov.it

Abstract. This study aimed to assess the predictive ability of genomic selection (GS) for biomass yield of alfalfa and pea, considering different data sets, GS models, and thresholds for genotype missing data. An additional aim was to briefly devise the incorporation of GS into breeding schemes of these crops. For alfalfa, the predictive ability of best GS models ranged from $r = 0.18$ to $r = 0.36$ in three data sets. The lowest value (observed in a data set with higher experimental error or lower genetic variation relative to the other data sets) may still be of practical interest, given the long selection cycle and the low narrow-sense heritability of biomass yield in this crop. For pea biomass yield, the predictive ability of best GS models averaged $r = 0.45$ across three recombinant inbred line (RIL) populations. Predictions were less accurate for this trait than for pea straw or grain yield. GS is a promising approach but its adoption implies important modifications of alfalfa and pea breeding schemes. We identified five stages of GS-based selection schemes, whose implementation depends largely on the reproductive system of the target species.

Keywords: Genomic selection accuracy · Forage yield · *Medicago sativa*
Pisum sativum · Predictive ability · Straw

1 Introduction

Socio-economic and environmental factors point to the pivotal role of legume-based crops in future crop-livestock systems of southern Europe (Annicchiarico 2017). Alfalfa (alias lucerne, *Medicago sativa* L.) is the most-grown perennial forage in this region. Its genetic variation includes germplasm with outstanding drought tolerance (Annicchiarico et al. 2011), which can be exploited for improving crop adaptation to the predicted adverse effects of climate change. Regional production of hay or silage may also rely on annual legumes, particularly in severely drought-prone environments where perennials may lack sufficient persistence. Recent findings have highlighted the interest of field pea (*Pisum sativum* L.) over vetch species (*Vicia* spp.) in this respect, both as a pure stand crop and for intercropping with cereals (Annicchiarico et al. 2017b). Maximizing the aerial biomass of semi-dwarf pea germplasm has crucial importance not only for pure stands but also for pea-cereal intercropping, where it ensures sufficient legume content and competitive ability against cereals (Annicchiarico et al. 2013).

© Springer International Publishing AG, part of Springer Nature 2018
G. Brazauskas et al. (Eds.): *Breeding Grasses and Protein Crops in the Era of Genomics*, pp. 259–264, 2018.
https://doi.org/10.1007/978-3-319-89578-9_47

Genomic selection (GS) pools phenotyping and genotyping data of a genotype sample representing a target genetic base (reference population) into a model that estimates breeding values for future plant selection (Heffner et al. 2009). GS has taken impulse from the development of genotyping-by-sequencing (GBS) (Elshire et al. 2011), which can produce thousands of genome-wide markers at a lower cost than SNP array platforms (albeit with large amounts of missing data). While predicting pure line performance is the obvious aim of GS in inbred species (such as pea), predicting the breeding value of candidate parent genotypes for synthetic varieties of outbred species (such as alfalfa) can be pursued by genotyping a set of parent genotypes and phenotyping their half-sib progenies (Annicchiarico et al. 2015a). First results for GS prediction of alfalfa forage yield or pea grain yield were promising (Annicchiarico et al. 2015b, 2017a; Li et al. 2015). Positive results emerged as well for prediction of some grain yield components of pea (Burstin et al. 2015).

This study pooled results for different material and/or cropping conditions with the aim to assess the predictive ability of GS for biomass yield of alfalfa and pea. An additional aim was to briefly devise the incorporation of GS into the breeding scheme of these crops.

2   Material and Methods

Alfalfa genotypes were phenotyped for dry biomass yield under dense-stand conditions of their half-sib progenies in three experiments termed hereafter as data sets. Data set 1 comprised 154 genotypes from a broadly-based population of Mediterranean germplasm, phenotyped under water-favourable conditions in a managed environment (750 mm of water over March–October) over four harvests of one year. Data set 2 included the same material, phenotyped under moderate drought stress in a managed environment (on average, 455 mm of water over March–October) over seven harvests across two years and the following spring. Data set 3 included 124 parent genotypes from a broadly-based population of landrace and variety germplasm from the Po Valley, phenotyped in Lodi (northern Italy) under field conditions and moderate drought stress (on average, 454 mm of rainfall plus irrigation water over March–October) over 12 harvests across two years and the following spring. Annicchiarico et al. (2015b) described procedures of GBS and SNP data calling for these data sets, as well as phenotyping procedures and results generated by seven GS models for the first and the third data set. This study adds original results relative to the second data set, whose experiment was carried out using same procedures (for plot size, experimental design, etc.) as the first data set but different drought stress level and experiment duration. For this data set, we exploited SNP data from Annicchiarico et al. (2015b) and the two GS models that proved more predictive for yield in the other two data sets, namely, Ridge Regression BLUP (rrBLUP), and Support Vector Regression using Linear Kernel (SVR-lin).

For pea, an earlier study (Annicchiarico et al. 2017a) reported the predictive ability of four GS models for grain yield under severe managed drought stress (120 mm of water over the period March–May) of three recombinant inbred line (RIL) populations, each including 105 lines. Here, we added information on GS predictive ability for dry biomass and straw yield assessed in the same experiment. The RILs were issued by connected crosses between three semi-dwarf cultivars (Attika; Isard; Kaspa) that exhibited high and stable grain yield across climatically-contrasting Italian sites (Annicchiarico 2005; Annicchiarico and Iannucci 2008). Attika and Kaspa displayed high biomass yield too, and proved suitable for forage production in mixed cropping with cereals (Annicchiarico et al. 2013, 2017b). We used GBS-based SNP data from Annicchiarico et al. (2017a) for the cautious minimum read depth of six for SNP genotype calling (given some heterozygosity expected in the genotyped F_6 generation), and adopted the two GS models that were more predictive for grain yield in that study, i.e., Bayesian Lasso (BL) and rrBLUP.

GS predictive ability (i.e., the correlation between GS-predicted values and observed values) was assessed across genotype SNP missing data thresholds for marker retention in the range 10–50%, using missing data imputation and cross-validation procedures described earlier (Annicchiarico et al. 2015b, 2017a). Pea GS models were trained over the three RIL populations without imputing genetic structure information, assessing their predicting ability on the single populations.

3   Results and Discussion

For unpublished results of alfalfa (second data set), the best GS configuration for predicting biomass yield was provided by the SVR-lin model with genotype missing data threshold of 40%, whose predictive ability reached $r = 0.18$ using 10911 polymorphic SNP markers. The rrBLUP model performed nearly as well ($r = 0.17$). Best predictions for this data set were distinctly lower than those observed for biomass yield of the same material under favourable cropping conditions (which featured distinctly lower experiment error CV, i.e., 14.1 *vs* 19.8%), or yield of a different reference population under moderate drought stress (which featured distinctly higher genetic variance CV, i.e., 22.8 *vs* 14.0%) (Table 1). It should be noted that even $r = 0.18$, although fairly unsatisfactory, could still provide a sizable advantage for GS over half-sib progeny based phenotypic selection in terms of predicted yield gains per unit time (Annicchiarico et al. 2015a, b). This would descend from assuming one year for each GS selection cycle and five years for each progeny-based phenotypic selection cycle, with narrow-sense heritability in the range 0.15–0.30 for biomass yield as indicated by various studies (Annicchiarico 2015).

According to pea GS results averaged across the three RIL populations, best predictions of aerial biomass were provided by the BL model with genotype missing rate of 30%. This configuration achieved moderately high predictive ability, i.e., $r = 0.45$, using 1537 polymorphic SNP markers over the three populations. However, predictive ability values were nearly identical in the range 20–50% of genotype missing data for the two GS models (data not shown). Best predictions for the single RIL populations ranged from 0.29 to 0.60. For straw yield, best predictions were provided by the BL model with 20% genotype missing rate, which displayed an average predicting ability of $r = 0.57$. Prediction for grain yield displayed the highest accuracy, averaging $r = 0.71$ (using BL with 20% missing genotype data).

We expected worse GS predictions for alfalfa than for pea RILs, owing to much shorter linkage disequilibrium and the impossibility to exploit non-additive genetic

Table 1. Predictive ability (PA; correlation between genome-based predicted values and observed values) of best genomic selection models for genotype breeding value of production traits, for two reference populations of alfalfa and three recombinant inbred line (RIL) populations of pea.

Crop	Population	Trait	Drought stress	PA	Source
Alfalfa	Po Valley	Biomass	Moderate	0.32	Annicchiarico et al. (2015b)
Alfalfa	Mediterranean	Biomass	Very limited	0.36	Annicchiarico et al. (2015b)
Alfalfa	Mediterranean	Biomass	Moderate	0.18	Unpublished data
Pea	Mean of 3 RIL pop.	Biomass	Severe	0.45	Unpublished data
Pea	Mean of 3 RIL pop.	Straw	Severe	0.57	Unpublished data
Pea	Mean of 3 RIL pop.	Grain	Severe	0.71	Annicchiarico et al. (2017a)

variation in half-sib progeny-based selection of alfalfa parents. However, GS provides greater opportunity for time reduction of selection cycles in a perennial such as alfalfa, justifying our interest even in low predictive ability values in this species.

Our results suggest that GS may already be convenient for breeding programs of alfalfa and pea. However, its incorporation would require important modifications of their selection schemes. This is summarized in Fig. 1, where five basic selection stages are identified whose implementation depends on the reproductive system of the target

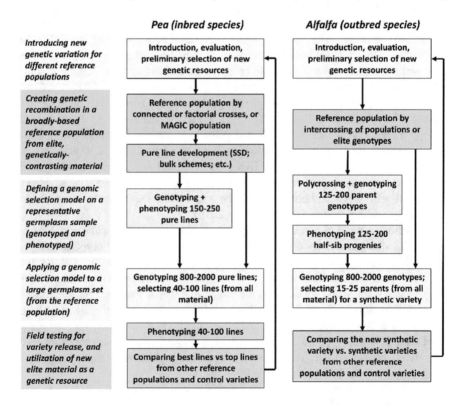

Fig. 1. Possible selection schemes integrating genomic selection, for pea and alfalfa

species (outbred or inbred). The inclusion of GS implies (i) the construction of one or more reference populations, (ii) the definition of one GS model for each target trait in each population using a genotype sample, (iii) the application of the model(s) to a wide set of genotypes from each population, and (iv) the final field test of a reduced set of GS-selected lines (inbreds) or the GS-selected synthetic variety (outbreds). For inbreds, the reference population may conveniently include a set of RILs with partly common ancestors (as here) to facilitate the definition of a common GS model, or one MAGIC population. Key scientific questions remains, *inter alia*, the ability of GS models built on one population to predict phenotypes of other populations, and the verification of actual yield gains obtained via GS.

Acknowledgments. This work is part of the EraNet-ARIMNet project 'Resilient, water- and energy-efficient forage and feed crops for Mediterranean agricultural systems (REFORMA)' funded for Italy by the Italian Ministry of Agriculture, Food and Forestry Policy. We are grateful to E.C. Brummer and B. Ferrari for scientific support, and S. Proietti, A. Passerini and P. Gaudenzi for valuable technical assistance.

References

Annicchiarico P (2005) Scelta varietale per pisello e favino rispetto all'ambiente e all'utilizzo. Informatore Agrario 61(49):47–52

Annicchiarico P (2015) Alfalfa forage yield and leaf/stem ratio: narrow-sense heritability, genetic correlation, and parent selection procedures. Euphytica 205:409–420

Annicchiarico P (2017) Feed legumes for truly sustainable crop-animal systems. Italian J Agron 12(2). https://doi.org/10.4081/ija.2017.880

Annicchiarico P, Iannucci A (2008) Adaptation strategy, germplasm type and adaptive traits for field pea improvement in Italy based on variety responses across climatically contrasting environments. Field Crops Res 108:133–142

Annicchiarico P, Barrett B, Brummer EC, Julier B, Marshall AH (2015a) Achievements and challenges in improving temperate perennial forage legumes. Crit Rev Plant Sci 34:327–380

Annicchiarico P, Nazzicari N, Li X, Wei Y, Pecetti L, Brummer EC (2015b) Accuracy of genomic selection for alfalfa biomass yield in different reference populations. BMC Genom 16:1020

Annicchiarico P, Nazzicari N, Pecetti L, Romani M, Ferrari B, Wei Y, Brummer EC (2017a) GBS-based genomic selection for pea grain yield under severe terminal drought. The Plant Genome 10(2). https://doi.org/10.3835/plantgenome2016.07.0072

Annicchiarico P, Pecetti L, Abdelguerfi A, Bouizgaren A, Carroni AM, Hayek T, Bouzina M, Mezni M (2011) Adaptation of landrace and variety germplasm and selection strategies for lucerne in the Mediterranean basin. Field Crops Res 120:283–291

Annicchiarico P, Ruda P, Sulas C, Pitzalis M, Salis M, Romani M, Carroni AM (2013) Optimal plant type of pea for mixed cropping with cereals. In: Barth S, Milbourne D (eds), Breeding Strategies for Sustainable Forage and Turf Grass Improvement. Springer, Dordrecht, pp 341–346

Annicchiarico P, Thami Alami I, Abbas K, Pecetti L, Melis RAM, Porqueddu C. (2017b) Performance of legume-based annual forage crops in three semi-arid Mediterranean environments. Crop Pasture Sci 8(10-11):932–941. https://doi.org/10.1071/CP17068

Burstin J, Salloignon P, Chabert-Martinello M, Magnin-Robert J-B, Siol M, Jacquin F, Chauveau A, Pont C, Aubert G, Delaitre C, Truntzer C, Duc G (2015) Genetic diversity and trait genomic prediction in a pea diversity panel. BMC Genom 16:105

Elshire RJ, Glaubitz JC, Sun Q, Poland JA, Kawamoto K, Buckler ES, Mitchell SE (2011) A robust, simple genotyping-by-sequencing (GBS) approach for high diversity species. PLoS ONE 6(5):e19379

Heffner EL, Sorrells ME, Jannink J-L (2009) Genomic selection for crop improvement. Crop Sci 49:1–12

Li X, Wei Y, Acharya A, Hansen JL, Crawford JL, Viands DR, Michaud R, Claessens A, Brummer EC (2015) Genomic prediction of biomass yield in two selection cycles of a tetraploid alfalfa breeding population. Plant Genome 8:2

Author Index

A

Aavola, R., 3, 186
Adamski, T., 51
Akinroluyo, O., 73
Aleliūnas, A., 3, 231, 254
Allen, D., 243
Alves, M. L., 118
Annicchiarico, P., 113, 259
Aper, J., 166
Armstead, I., 243
Ashikaga, K., 125
Asp, T., 3, 145, 160
Augustyniak, A., 176
Ayling, S., 20

B

Bachmann-Pfabe, S., 27
Baert, J., 78, 155, 166
Balsevičius, A., 56
Barre, P., 14, 103, 226, 249
Barrett, B., 203
Barth, S., 155
Bartoš, J., 155
Barzyk, P., 131
Belanche, A., 121
Beļeviča, V., 62
Bender, A., 186
Bertrand, A., 209
Bērziņš, P., 62
Boller, B., 33
Brady, J., 46
Brazauskas, G., 3, 231
Bronze, M. R., 118
Butkutė, B., 56, 83
Byrne, S., 160

C

Carroni, A. M., 113
Černoch, V., 155
Chebotar, G., 220
Chebotar, S., 220
Claessens, A., 209
Cnops, G., 39
Cougnon, M., 78

D

De Ron, A. M., 9
De Vega, J., 20
Dehmer, K. J., 14, 27
Delaunay, S., 249
Doležel, J., 155
Dubeux Jr., J. C., 46

E

Ergon, Å., 109
Escobar-Gutiérrez, A., 14
Evans, C., 243

F

Felder, T., 33
Ferreyra, M. J., 9
Flajoulot, S., 226
Flay, C., 203
Forrest, F. S., 46
Fourie, J., 203
Fujii, H., 125

G

Gailīte, A., 62
Gasior, D., 243
Gebbie, S., 203

Gensollen, V., 249
Ghamkhar, K., 203
Ghesquiere, A., 166
Ghesquière, M., 145, 155, 196
Górynowicz, B., 191
Gréard, C., 226
Grogan, D., 155
Gylstrøm, K. H., 3

H
Hagedorn, M., 203
Harper, J., 155, 243
Hartmann, S., 171
Hegarty, M., 20
Helgadottir, A., 3
Hentrup, S., 160
Hersemann, L., 150
Hsiao, J., 203
Humphreys, M., 121, 155, 243
Huyghe, C., 113

I
Irie, K., 203
Isolahti, M., 3
Ivanyuk, S., 220

J
Jansons, A., 62
Jonavičienė, K., 254
Jones, C., 20
Julier, B., 103, 226, 249

K
Kaczmarek, Z., 51
Kamau, S., 121
Kamel, K., 131
Karagić, Đ., 103
Kemešytė, V., 56, 73, 88, 96
Kingston-Smith, A., 121
Knorst, V., 150
Knot, P., 137
Knotová, D., 137, 237
Kölliker, R., 150
Kopecký, D., 33, 155
Korniychuk, O., 220
Kosmala, A., 176
Kovi, M. R., 3, 109
Kristjánsdóttir, T. A., 3
Kroc, M., 131
Krystkowiak, K., 51
Książczyk, T., 155
Kuczyńska, A., 51

L
Lafaillette, F., 249
Lambroni, P., 249
Lang, J., 237
Larsen, A. S., 3
Leitão, S. T., 118
Lemežienė, N., 83, 88
Leyva, M. O., 160
Liatukienė, A., 67
Lloyd, D., 20, 103

M
Manunza, P., 113
Marum, P., 3
Masajada, K., 176
Mathews, R., 243
Mecha, E., 118
Menéndez-Sevillano, M. C., 9
Mikaliūnienė, J., 96
Mikołajczak, K., 51
Mikulski, W., 131, 191
Milvang, Ø. W., 109
Milvang, O., 109
Mladenović-Drinić, S., 215
Muir, J. P., 46
Muylle, H., 14

N
Nagy, I., 160
Nazzicari, N., 259
Nedělník, J., 181
Nikolić, A., 215
Norkevičienė, E., 56, 83, 88, 96

O
Ogrodowicz, P., 51
Oppermann, M., 27
Østrem, L., 145, 155

P
Padarauskas, A., 83
Paina, C., 3
Paszkowski, E., 155
Pawłowicz, I., 176
Pecetti, L., 113, 259
Pelikán, J., 137
Pereira, A. B., 118
Perić, V., 215
Perlikowski, D., 176
Persson, C., 3
Petrauskas, G., 56, 96
Pilarczyk, W., 191

Pitman, W. D., 46
Pose, J. D., 9
Powell, W., 20

R
Rancāne, S., 3, 62
Ratajczak, D., 176
Reheul, D., 78
Rocher, S., 209
Rognli, O. A., 3, 145
Roldán-Ruiz, I., 39, 103
Romani, M., 259
Roschanski, A. M., 14
Ruņģis, D., 62
Rybiński, W., 131

S
Sampoux, J. P., 14
Sanada, Y., 145
Santoni, S., 226
Schäufele, R., 171
Skládanka, J., 137
Skøt, L., 20, 103
Sokolovič, D., 155
Sooväli, P., 186
Srebrić, M., 215
Statkevičiūtė, G., 73, 96, 254
Stesele, V., 62
Stewart, A., 203
Stopyra, P., 51
Studer, B., 150
Stukonis, V., 88
Surma, M., 51

Święcicki, W. K., 51
Święcicki, W., 131, 191

T
Tamm, S., 186
Tamura, K., 125
Tanaka, T., 125
Taujenis, L., 83
Thomas, A., 243
Thomas, I., 14
Thomasset, M., 249
Trněný, O., 181

V
Vaz Patto, M. C., 118
Vēzis, I., 62
Vleugels, T., 39
Vymyslický, T., 237

W
Weise, S., 27, 103
Werner, A., 203
Westermeier, P., 171
Widmer, F., 150
Wilczura, P., 131
Willner, E., 14, 27

Y
Yamada, T., 125

Z
Zharikova, D., 220
Zwierzykowski, W., 176
Zwierzykowski, Z., 155

Printed in the United States
By Bookmasters